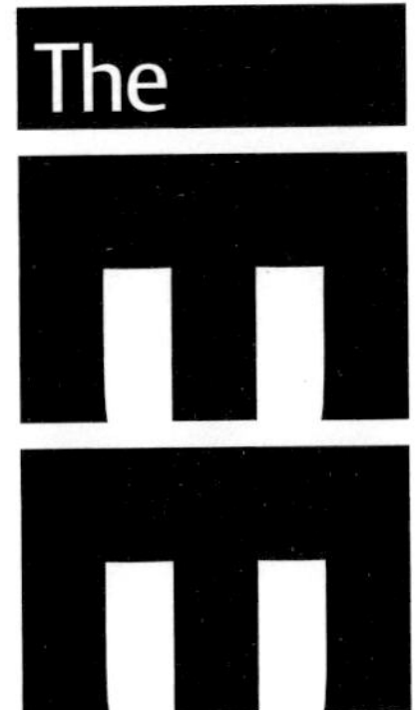

Guidance Note 5

Protection Against Electric Shock

17th IEE Wiring Regulations Seventeenth Edition
BS 7671:2008 Requirements for Electrical Installations

Published by The Institution of Engineering and Technology, London, United Kingdom

The Institution of Engineering and Technology is registered as a Charity in England & Wales (no. 211014) and Scotland (no. SC038698).

The Institution of Engineering and Technology is the new institution formed by the joining together of the IEE (The Institution of Electrical Engineers) and the IIE (The Institution of Incorporated Engineers). The new Institution is the inheritor of the IEE brand and all its products and services, such as this one, which we hope you will find useful. The IEE is a registered trademark of the Institution of Engineering and Technology.

First published 1992 (0 85296 538 9)
Reprinted (with minor amendments) 1993
Second edition (incorporating Amendment No. 1 to BS 7671:1992) 1995 (0 85296 869 8)
Third edition (incorporating Amendment No. 2 to BS 7671:1992) 1999 (0 85296 958 9)
Fourth edition (incorporating Amendment No. 1 to BS 7671:2001) 2003 (0 85296 993 7)
Reprinted (incorporating Amendment No. 2 to BS 7671:2001) 2004
Fifth edition (incorporating BS 7671:2008) 2009 (978-0-86341-859-4)

Copies of this publication may be obtained from:
The Institution of Engineering and Technology
PO Box 96
Stevenage
SG1 2SD, UK
Tel: +44 (0)1438 767328
Email: sales@theiet.org
www.theiet.org/publishing/books/wir-reg/

ISBN 978-0-86341-859-4

Typeset in the UK by Carnegie Book Production, Lancaster
Printed in the UK by Printwright Ltd, Ipswich

Contents

Cooperating organisations

The Institution of Engineering and Technology acknowledges the contribution made by the following organisations in the preparation of this Guidance Note.

Association of Manufacturers of Domestic Appliances
S.A. MacConnacher BSc CEng MIEE

BEAMA Installation Ltd
Eur Ing M.H. Mullins BA CEng FIEE FIIE
P. Sayer IEng MIET GCGI

British Cables Association
C.K. Reed IEng MIET

British Electrotechnical & Allied Manufacturers Association Ltd
P.D. Galbraith IEng MIET

British Standards Institution

City & Guilds of London Institute
H.R. Lovegrove IEng FIET

Electrical Contractors' Association
D. Locke BEng(Hons) CEng MIEE MIEEE ACIBSE

Electrical Contractors' Association of Scotland (SELECT)
D. Millar IEng MIIE MILE

Electrical Safety Council
S.J. Hesketh IEng MIET

GAMBICA Association Ltd
M. Hadley

Health and Safety Executive
K. Morton BSc CEng MIEE

Institution of Engineering and Technology
P.R.L. Cook CEng FIET MCIBSE (Editor)
G.D. Cronshaw IEng FIET
P.E. Donnachie BSc CEng FIET
Eur Ing L. Markwell MSc BSc CEng MIET MCIBSE LCGI

Lighting Association
L.C. Barling

Society of Electrical and Mechanical Engineers serving Local Government
C.J. Tanswell CEng MIET MCIBSE

Acknowledgements

References to British Standards, CENELEC Harmonization Documents and International Electrotechnical Commission standards are made with the kind permission of BSI. Complete copies can be obtained by post from:

BSI Customer Services
389 Chiswick High Road
London W4 4AL
Tel: +44 (0)20 8996 9001
Fax: +44 (0)20 8996 7001
Email: orders@bsi-global.com

BSI also maintains stocks of international and foreign standards, with many English translations. Up-to-date information on BSI standards can be obtained from the BSI website: www.bsi-global.com

Advice is included from Engineering Recommendation G12/3 'Requirements for the application of protective multiple earthing to low voltage networks' with the kind permission of the Energy Networks Association Limited. Complete copies of this and other Engineering Recommendations can be obtained by post from:

Energy Networks Association Ltd
6th Floor, Dean Bradley House
52 Horseferry Road
London SW1P 2AF
Tel: +44 (0)20 7706 5100
Email: richard.legros@energynetworks.org

Documents available from their website www.energynetworks.org include Technical Specifications, BEBS Specifications, Engineering Recommendations and a variety of reports.

Most of the illustrations within this publication were provided by Rod Farquhar Design: www.rodfarquhar.co.uk

Cover design and illustration were created by The Page Design: www.thepagedesign.co.uk

Preface

This Guidance Note is part of a series issued by the Institution of Engineering and Technology to explain and enlarge upon the requirements in BS 7671:2008, the 17th Edition of the IEE Wiring Regulations.

Note that this Guidance Note does not ensure compliance with BS 7671. It is intended to explain some of the requirements of BS 7671 but readers should always consult BS 7671 to satisfy themselves of compliance.

The scope generally follows that of BS 7671; the relevant Regulations and Appendices are noted in the margin. Some Guidance Notes also contain material not included in BS 7671:2008 but which was included in earlier editions of the Wiring Regulations. All of the Guidance Notes contain references to other relevant sources of information.

Electrical installations in the United Kingdom that comply with BS 7671 are likely to satisfy Statutory Regulations such as the Electricity at Work Regulations 1989; however, this cannot be guaranteed. It is stressed that it is essential to establish which Statutory and other Regulations apply and to install accordingly. For example, an installation in premises subject to licensing may have requirements different from, or additional to, BS 7671 and these will take precedence.

Additionally, the requirements of the person ordering the work should be met, unless there is conflict with the requirements of BS 7671 or the Electricity at Work Regulations. 110.1

Users of this Guidance Note should assure themselves that they have complied with any legislation that post-dates the publication.

Introduction

This Guidance Note is principally concerned with Chapter 41 of BS 7671 'Protection against electric shock'. Other relevant parts of BS 7671 are also considered.

Neither BS 7671 nor the Guidance Notes are design guides. It is essential to prepare a full design and specification prior to commencement or alteration of an electrical installation.

The design and specification should set out the requirements and provide sufficient information to enable competent persons to carry out the installation and to commission it. The specification must include a description of how the system is to operate and all the design and operational parameters. It must provide for all the commissioning procedures that will be required and for the provision of adequate information to the
user. This should be by means of an operation and maintenance manual or schedule, 514.9
and 'as fitted' drawings if necessary.

It must be noted that it is a matter of contract as to which person or organisation is responsible for the production of the parts of the design, specification, construction and verification of the installation and any operational information.

The persons or organisations who may be concerned in the preparation of the works include:

The Designer
The CDM Coordinator
The Installer
The Distributor of Electricity
The Installation Owner (Client) and/or User
The Architect
The Fire Prevention Officer
All Regulatory Authorities
Any Licensing Authority
The Health and Safety Executive

In producing the design, advice should be sought from the installation owner and/or 132.7
user as to the intended use. Often, as in a speculative building, the intended use is unknown. The specification and/or the operational manual must set out the basis of use for which the installation is suitable.

Precise details of each item of equipment should be obtained from the manufacturer 133.1
and/or supplier and compliance with appropriate standards confirmed. Sect 511

The operational manual must include a description of how the system as installed is to operate and all commissioning records. The manual should also include manufacturers' technical data for all items of switchgear, luminaires, accessories, etc. and any special instructions that may be needed.

The Health and Safety at Work etc. Act 1974 Section 6 and the Construction (Design and Management) Regulations 2007 are concerned with the provision of information, and guidance on the preparation of technical manuals is given in the BS 4884 series *Technical manuals* and the BS 4940 series *Technical information on constructional products and services*. The size and complexity of the installation will dictate the nature and extent of the manual.

The order in which the protective measures are considered in this Guidance Note is significant in that the protective measure of automatic disconnection of supply is the measure that is employed in almost every electrical installation.

Statutory regulations and the wiring regulations 1

1.1 Statutory Regulations

Statutory legislation prescribing requirements for electrical safety and protection against electric shock includes the Health and Safety at Work etc. Act 1974, and in particular the Electricity at Work Regulations 1989 (EAWR), the Provision and Use of Work Equipment Regulations 1998 (PUWER), the Workplace (Health, Safety and Welfare) Regulations 1992, the Electricity Safety, Quality and Continuity Regulations 2002 (ESQCR), and the Building Regulations.

1.2 The Electricity at Work Regulations 1989 as amended

The Electricity at Work Regulations commented on in Table 1.1 are of particular relevance to the provision of suitable means of protection against electric shock in the design and construction of an installation.

▼ **Table 1.1** Electricity at Work Regulations of particular relevance to the provision of suitable means of protection against electric shock in the design and construction of an installation

EAWR Regulation	Commentary
4(1)	Addresses the construction of systems. An installation is part of a system, and the Regulation embraces the arrangement of its components, including the need for suitable electrical protective devices and other means to prevent danger from electric shock.
5	Requires the strength and capability of electrical equipment not to be exceeded and thereby give rise to danger. In the context of protection against electric shock, insulation must remain effective under normal and any likely transient overvoltage conditions. Protective conductors must be adequately rated to survive earth fault conditions and allow satisfactory operation of relevant protective devices.
7	Requires specified protective measures to be applied, or other suitable precautions to be taken, to prevent contact with live parts.
8	Requires suitable precautions to be taken to prevent danger arising from faults in the installation.
9	Requires the integrity of any referenced conductor, e.g. an earthed neutral or combined neutral and protective conductor, to be ensured to prevent danger.
10	Requires, where necessary to prevent danger, every joint and connection to be mechanically and electrically suitable for use. Joints and connections in live conductors must be properly insulated and capable of safely withstanding any likely fault conditions. Joints and connections in protective conductors (see also Regulation 5 above) must be made at least as carefully as those in live conductors.
12	Requires means to be available for cutting off the supply of electrical energy to, and for isolation of, any electrical equipment.

As might be expected, the 'fundamental principles' of BS 7671 (Chapter 13) and those of the Electricity at Work Regulations which are concerned with electrical installations are worded similarly. It is intended that, by meeting the technical requirements set out in BS 7671, compliance with the fundamental principles of BS 7671 and also of the relevant provisions of Statutory Regulations is achieved.

1.3 The Electricity Safety, Quality and Continuity Regulations 2002 as amended

There are a number of references to BS 7671 in the Electricity Safety, Quality and Continuity Regulations 2002 (ESQCR).

Requirements in the ESQCR that are concerned with the safety of low voltage electrical installations include those listed in Table 1.2.

▼ **Table 1.2** Electricity Safety, Quality and Continuity Regulations of particular relevance to the provision of suitable means of protection against electric shock in the supply to an installation

ESQCR Regulation	Commentary
3(1)(b)	Regulation 3(1) requires, amongst other things, that distributors and meter operators ensure that their equipment is protected (both electrically and mechanically) so as to prevent danger.
8(3) and 8(4)	The requirements of Regulation 8(3) are that a distributor shall ensure – (a) the outer conductor of any electric line which has concentric conductors is connected with earth; (b) every supply neutral conductor is connected with earth at, or as near as is reasonably practicable to, the source of voltage (with a permitted exception); and (c) no impedance is inserted in any connection with earth of a low voltage network other than that required for the operation of switching devices or of instruments or equipment for control, telemetry or metering. 8(4) A consumer shall not combine the neutral and protective functions in a single conductor in his consumer's installation.
9(2), 9(3) and 9(4)	*Protective multiple earthing* (2) In addition to the neutral with earth connection required under Regulation 8(3) a distributor shall ensure that the supply neutral conductor is connected with earth at – (a) a point no closer to the distributor's source of voltage (as measured along the distributing main) than the junction between that distributing main and the service line which is most remote from the source; and (b) such other points as may be necessary to prevent, so far as is reasonably practicable, the risk of danger arising from the supply neutral conductor becoming open circuit. (3) Paragraph (2)(a) shall only apply where the supply neutral conductor of the service line referred to in paragraph (2)(a) is connected to the protective conductor of a consumer's installation. (4) The distributor shall not connect his combined neutral and protective conductor to any metalwork in a caravan or boat.

▼ **Table 1.2** *continued*

ESQCR Regulation	Commentary
24(1), 24(2), 24(3), 24(4) and 24(5)	(1) A distributor or meter operator shall ensure that each item of his equipment which is on a consumer's premises but which is not under the control of the consumer (whether forming part of the consumer's installation or not) is – (a) suitable for its purpose; (b) installed and, so far as is reasonably practicable, maintained so as to prevent danger; and (c) protected by a suitable fusible cut-out or circuit breaker which is situated as close as is reasonably practicable to the supply terminals. (2) Every circuit breaker or cut-out fuse forming part of the fusible cut-out mentioned in paragraph (1)(c) shall be enclosed in a locked or sealed container as appropriate. (3) Where they form part of his equipment which is on a consumer's premises but which is not under the control of the consumer, a distributor or meter operator (as appropriate) shall mark permanently, so as clearly to identify the polarity of each of them, the separate conductors of low voltage electric lines which are connected to supply terminals and such markings shall be made at a point which is as close as is practicable to the supply terminals in question. (4) Unless he can reasonably conclude that it is inappropriate for reasons of safety, a distributor shall, when providing a new connection at low voltage, make available his supply neutral conductor or, if appropriate, the protective conductor of his network for connection to the protective conductor of the consumer's installation. (5) In this regulation the expression 'new connection' means the first electric line, or the replacement of an existing electric line, to one or more consumer's installations.
25(2)	Regulation 25(2) requires the distributor to refuse connection of the supply if evidence of compliance with BS 7671 is not provided.
27(1), 27(2), 27(3)	(1) Before commencing a supply to a consumer's installation, or when the existing supply characteristics have been modified, the supplier shall ascertain from the distributor and then declare to the consumer – (a) the number of phases; (b) the frequency; and (c) the voltage, at which it is proposed to supply electricity and the extent of the permitted variations thereto. (2) Unless otherwise agreed in writing between the distributor, the supplier and the consumer (and if necessary between the distributor and any other distributor likely to be affected) the frequency declared pursuant to paragraph (1) shall be 50 hertz and the voltage declared in respect of a low voltage supply shall be 230 volts between the phase and neutral conductors at the supply terminals. (3) For the purposes of this regulation, unless otherwise agreed in writing by those persons specified in paragraph (2), the permitted variations are – (a) a variation not exceeding 1 per cent above or below the declared frequency; (b) in the case of a low voltage supply, a variation not exceeding 10 per cent above or 6 per cent below the declared voltage at the declared frequency;
28	A distributor shall provide, in respect of any existing or proposed consumer's installation which is connected or is to be connected to his network, to any person who can show a reasonable cause for requiring the information, a written statement of – (a) the maximum prospective short circuit current at the supply terminals; (b) for low voltage connections, the maximum earth loop impedance of the earth fault path outside the installation; (c) the type and rating of the distributor's protective device or devices nearest to the supply terminals; (d) the type of earthing system applicable to the connection; and (e) the information specified in regulation 27(1), which apply, or will apply, to that installation.

1.4 The Building Regulations

1.4.1 England and Wales

The Building Regulations 2000 now include requirements for electrical installations, in particular in Part P. For the purposes of providing practical guidance with respect to the requirements of the Building Regulations for England and Wales, the Secretary of State has issued a series of Approved Documents including Part P Electrical safety. Persons responsible for work within the scope of Part P of the Building Regulations may also be responsible for ensuring other parts of the Building Regulations where relevant are complied with, particularly if there are no other parties involved with the work.

The Building Regulations (Regulation 4(2)) require that, on completion of the work, the building should be no worse in terms of the level of compliance with the other applicable Parts of Schedule 1 to the Building Regulations, including Parts A, B, C, E, F, L and M.

Part P applies to electrical installations in buildings or parts of buildings comprising:

1 dwellinghouses and flats;
2 dwellings and business premises that have a common metered supply – for example shops and public houses with a flat above with a common meter;
3 common access areas in blocks of flats such as corridors and staircases;
4 shared amenities of blocks of flats such as laundries and gymnasiums.

Part P applies also to parts of the above electrical installations:

5 in or on land associated with the buildings – for example, Part P applies to fixed lighting and pond pumps in gardens;
6 in outbuildings such as sheds, detached garages and greenhouses.

Chap 13 The requirements of Part P should be met by adherence to the fundamental principles for achieving safety given in BS 7671, Chapter 13. To achieve these requirements electrical installations in dwellings, etc. must be designed and installed to afford appropriate protection against mechanical and thermal damage, and so that they do not present electric shock and fire hazards to people, and be suitably inspected and tested to verify that they meet the relevant equipment and installation standards.

Guidance is given in the IET publication *Electrician's Guide to the Building Regulations*.

1.4.2 Scotland

Appx 2 The Building (Scotland) Regulations 2004 include requirements to secure the health, safety, welfare and convenience of persons in or about buildings and of others who may be affected by buildings or matters connected with buildings; to further the conservation of fuel and power; and to further the achievement of sustainable development.

Requirements for electrical installations in Scotland are addressed by standard 4.5 – electrical safety for all buildings – and standard 4.6 – electrical fixtures for domestic buildings only – and, as in England and Wales, persons carrying out electrical installations must ensure that the work they carry out both complies with building regulations and the relevant functional standards.

There are no significant differences in general installation requirements for electrical work, with both Scotland and England & Wales citing BS 7671 (as amended) as the recommended means of satisfying building standards requirements. (See chapter 12 of the IET publication *Electrician's Guide to the Building Regulations*.)

1.4.3 Northern Ireland

The Department of Finance and Personnel is responsible for the Building Regulations (Northern Ireland) 2000. These contain at publication no particular requirements for the safety of electrical installations. Compliance with BS 7671 would be expected. Appx 2

Electric shock and protective measures 2

2.1 Electric shock

Electric shock is defined as *a dangerous physiological effect resulting from the* Part 2
passage of an electric current through a human body or livestock. The application of 131.2
the shock protection requirements of BS 7671 can be expected to give such protection to the average person, or, in other words, to most persons under most conditions. Requirements such as short disconnection times and low residual operating currents for residual current devices cannot be guaranteed to prevent death or serious injury due to electric shock in all circumstances. Some persons, such as the young or the elderly, may be more at risk. A useful reference document is IEC publication DD IEC/TS 60479-1:2005 *Effects of current on human beings and livestock*. This relates the magnitude and duration of electric currents to the probable severity of their effects, based on information gathered internationally.

The fundamental rule of protection against electric shock, according to BS EN 61140, Sect 410
is that hazardous-live-parts, such as energized conductors, must not be accessible and accessible conductive parts, such as metal enclosures of equipment or metal pipes, must not be hazardous-live, either in use without a fault (normal conditions) or under single-fault conditions, such as a fault from a live conductor to metal casing.

Protection under normal conditions is achieved by basic protection, formerly known as protection against direct contact, and protection under single-fault conditions is achieved by fault protection. Protection under earth fault conditions was previously referred to as protection against indirect contact.

In locations where the risk of electric shock is increased, for example due to a reduction Sect 700
in body resistance (due to water or sweating) or due to contact with Earth (working outdoors), BS 7671 prescribes supplementary or modified requirements such as:

- prohibition of the use of the less common protective measures such as obstacles and placing out of reach
- the requirement that only the protective measure of SELV is permitted
- a reduction in the nominal voltage of the SELV system
- more onerous requirements on ingress protection
- RCD protection.

2.2 Physiology of electric shock

IEC publication DD IEC/TS 60479-1:2005 *Effects of current on human beings and* 131.2
livestock (also published as BS PD 6519-1) advises that for currents less than 5 mA passing through the human body, there are usually no harmful physiological effects. As the body currents increase, so the risk of organic damage and probability of ventricular fibrillation increases, as can be seen in Figure 2.1 and Table 2.1.

(The philosophy of shock protection is discussed more fully in IET publication *Commentary on the IEE Wiring Regulations*.)

▼ **Figure 2.1** Conventional time/current zones of effects of a.c. currents (15 Hz to 100 Hz) on persons for a current path corresponding to left hand to feet (for explanation see Table 2.1) from DD IEC/TS 60479-1:2005 Figure 20

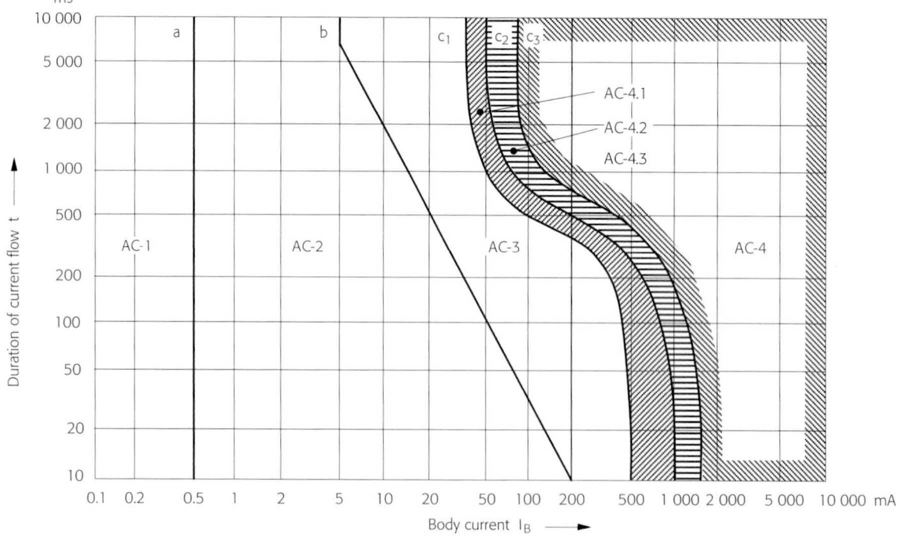

▼ **Table 2.1** The effects of current on the human body (Table 4 of IEC 479-1)

Zone designation	Zone limits	Physiological effects
AC-1	Up to 0.5 mA (curve a)	Perception possible but usually no 'startled' reaction
AC-2	0.5 mA up to curve b	Perception and involuntary muscular contractions likely but usually no harmful electrical physiological effects
AC-3	Curve b and above	Strong involuntary muscular contractions. Difficulty in breathing. Reversible disturbances of heart function. Immobilisation may occur. Effects increasing with current magnitude. Usually no organic damage to be expected
AC-4	Above curve c1	Patho-physiological effects may occur such as cardiac arrest, breathing arrest, and burns or other cellular damage. Probability of ventricular fibrillation increasing with current magnitude and time
AC-4.1	Between c1 and c2	Probability of ventricular fibrillation increasing up to about 5%
AC-4.2	Between c2 and c3	Probability of ventricular fibrillation up to about 50%
AC-4.3	Beyond curve c3	Probability of ventricular fibrillation above 50%

Note: For durations of current flow below 200 ms, ventricular fibrillation is only initiated within the vulnerable period if the relevant thresholds are surpassed. As regards ventricular fibrillation, Figure 2.1 relates to the effects of current which flows in the path left hand to feet. For other current paths, the heart current factor has to be considered.

2.3 Protective provisions and protective measures

BS 7671 requires two lines of defence against electric shock: a basic protective provision (e.g. basic insulation of live parts) and one or more fault protective provisions (e.g. supplementary insulation); see Figure 2.2. The combination of a basic protective provision (e.g. basic insulation of live parts) and one or more fault protective provisions is a protective measure (e.g. double insulation). Alternatively protection can be provided by an enhanced protective provision such as reinforced insulation; see Table 2.2. 131.2 410.3.2

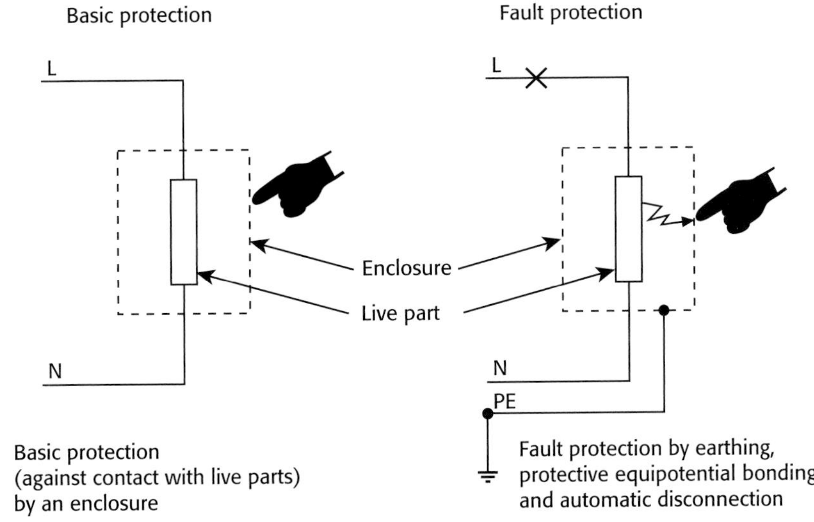

▼ **Figure 2.2**
Examples of basic protection and fault protection

In electrical installations the most commonly used protective measure is automatic disconnection of supply (previously called Earthed Equipotential Bonding and Automatic Disconnection of Supply (EEBADS), though this related only to fault protection). 410.3.3 Note

The protective measure may be required to be supplemented by an additional protective measure, such as a 30 mA RCD or supplementary bonding. 410.3.4

The detailed provisions for basic and fault protection are set out in a number of chapters and sections of BS 7671, for example:

- Chapter 13 Fundamental principles
- Chapter 41 Protection against electric shock
- Chapter 54 Earthing arrangements and protective conductors
- Section 701 Bathrooms.

Table 2.2 Summary of protective measures and protective provisions

	Protective measure	Protective provisions	
		Basic protection by	Fault protection by
Sect 411	Automatic disconnection of supply	Insulation of live parts, or barriers or enclosures	Protective earthing, protective equipotential bonding, automatic disconnection
Sect 412	Double insulation	Basic insulation	Supplementary insulation
	Reinforced insulation	Reinforced insulation	
Sect 413	Electrical separation for one item of equipment	Insulation of live parts	One item of equipment; simple separation from other circuits and Earth
Sect 414	Extra-low voltage (SELV and PELV)	Limitation of voltage, protective separation, basic insulation	
	For supervised installations:		
Sect 417	Obstacles	Obstacles	None
Sect 417	Placing out of reach	Placing out of reach	None
418.1	Non-conducting location	Insulation of live parts, or barriers or enclosures	No protective conductor; insulating floor and walls, spacings/obstacles between exposed-conductive-parts and extraneous-conductive-parts
418.2	Earth-free protective bonding	Insulation of live parts, or barriers or enclosures	Protective bonding, notices, etc.
418.3	Electrical separation for more than one item of equipment	Insulation of live parts	Simple separation from other circuits and Earth, non-earthed protective bonding, etc.

Provisions for basic protection 3

3.1 Introduction

Basic protection, or protection against direct contact with live parts, provides protection 411.2
under normal conditions and is most often provided by insulation and/or barriers or enclosures. Other protective measures are *obstacles* and *placing out of reach*, which are discussed in Chapter 9.

Table 3.1 summarises the purpose and scope of application of the above-mentioned protective measures.

▼ **Table 3.1** Provisions for basic protection

Provision (protection by)	Purpose	Application	
Insulation of live parts	Prevent contact with live parts	General	416.1
Barriers or enclosures	Prevent contact with live parts	General	416.2
Obstacles	Prevent unintentional bodily approach and unintentional contact with live parts	Limited to areas where access is restricted to skilled persons, or instructed persons under the supervision of skilled persons	417.2
Placing out of reach	Prevent unintentional contact with live parts	Limited to areas where access is restricted to skilled persons, or instructed persons under the supervision of skilled persons	417.3

3.2 Insulation

Insulation when provided for basic protection must completely cover the live parts. 416.1
The insulation must only be able to be removed by destruction, for example, stripping the insulation from an insulated conductor prior to connecting the conductor to a terminal. The measure is widely used in most installations, and the basic insulation covering on, for example, conductors of cables made to British Standards or equivalent standards affords adequate basic protection, providing the quality and effectiveness of the insulation is commensurate with the voltage to be applied to the conductors and other conditions of use, e.g. constant flexing in the case of a flexible cable.

Wherever possible, the application of insulation on site should be avoided by employing proper factory made and factory tested products and equipment. Where it becomes necessary to apply basic insulation to live parts during the erection of the installation, the quality of that insulation must be verified. This will normally require the use of high voltage test equipment and special test methods.

Some electrical equipment contains live parts which have paint, varnish, lacquer or a similar product applied over them, e.g. windings of a motor. Such finishes alone are not considered adequate insulation for basic protection, therefore live parts treated with any of these must not be accessible through apertures in the enclosure of the equipment concerned, e.g. IP2X.

132.7 One of the fundamental principles relates to the type of wiring and method of
521.10.1 installation including consideration of mechanical damage. An example of this is the requirement that non-sheathed cables for fixed wiring shall be enclosed in conduit, ducting or trunking.

522.6 Further mechanical protection may also be required for cables which are installed under a floor or above a ceiling, or within a wall or partition. Where required, the protection should provide a barrier to prevent the possible penetration of the cable insulation (including any insulating sheath) by a tool or fixing.

134.1.1 During erection, care must be taken to ensure that the effectiveness of insulation is not impaired through rough handling or poor workmanship (such as the removal of too much insulation from conductors at terminations or connections). Precautions are often required to prevent damage by other trades, e.g. to the insulation or sheathing of
522.1.2 cables during plastering. Thermoplastic (PVC) insulated cables should not be handled at ambient temperatures below 5 °C, or their insulation will be liable to cracking.

612.3 An insulation resistance test is required on completion of the fixed wiring of every installation, to verify that the measure is effective and complies with BS 7671.

3.3 Barriers or enclosures

Barriers or enclosures provided for basic protection must:

416.2.1
416.2.4
1. provide a degree of protection of at least IPXXB or IP2X, with permitted exceptions for access and replacement of parts

416.2.2
2. provide a degree of protection of IPXXD or IP4X for any readily accessible horizontal top surface

416.2.3
3. be firmly secured in place and have sufficient stability and durability taking into account the external conditions.

Barriers:

Part 2 A barrier is intended to provide protection against contact with live parts from any usual direction of access. It need not, for instance, have a top surface if access from above
416.2.4 that barrier is unlikely. A barrier may also be removable, for example to allow easy access to live parts, in which case one or more requirements specified by BS 7671 must be satisfied (see below).

Enclosures:

Part 2 An enclosure is employed to provide protection against contact with live parts from any direction.

For household installations, barriers or enclosures are an integral part of wiring accessories, consumer units and current-using equipment. This is also true of industrial installations where such features, together with obstacles, may well be specially designed and erected in order to obtain the protective measure.

The protection afforded by the enclosure of virtually any type of electrical equipment may be specified in accordance with the International Protection (IP) Code classification system described in BS EN 60529 *Specification for degrees of protection provided by enclosures* (IP Code). For basic protection the minimum degree of protection required by BS 7671 is IP2X (or IPXXB), and, in addition, any readily accessible horizontal top surface must provide a degree of protection of at least IPXXD or IP4X. (See Appendix B of Guidance Note 1 for information on the IP Code.) 416.2.1 416.2.2 GN1 Appx B

Where it is necessary to remove a barrier or open an enclosure or remove parts of enclosures, this must be possible only: 416.2.4

1. by the use of a key or tool, or
2. after disconnection of the supply to live parts against which the barriers or enclosures afford protection, restoration of the supply being possible only after replacement or reclosure of the barrier or enclosure, or
3. where an intermediate barrier providing a degree of protection of at least IPXXB or IP2X prevents contact with live parts, by the use of a key or tool to remove the intermediate barrier.

The requirements do not apply to a ceiling rose complying with BS 67, a cord operated switch complying with BS 3676, a bayonet lampholder complying with BS EN 61184 or an Edison screw lampholder complying with BS EN 60238.

Detailed construction requirements for low voltage switchgear and controlgear assemblies, including distribution boards, are given in the relevant parts of BS EN 60439. This standard covers all low voltage assemblies, both type-tested (TTA) and partially type-tested (PTTA). To comply with the standard, all external surfaces must conform to the minimum degree of protection stated above, i.e. IP2X. Most applications of barriers are also to be found in the construction of switchgear, etc. BS 7671 permits an opening larger than IP2X in a barrier or an enclosure, for functional reasons or to replace a part. For this purpose the opening must be no larger than is strictly necessary and the cover or barrier covering the opening should only be removable by the use of a tool and be fitted with a suitable permanent warning label. In this connection a coin is not a tool and such covers, etc. should not have fixings which can be undone by a coin. 416.2.1

BS EN 60439-1 also takes account of those situations where it is necessary, for reasons of operation, to gain access to the interior of an assembly whilst it is still live. Four forms of internal separation of circuits by barriers or partitions fitted within an assembly are specified in clause 7.7 of the standard.

When access to the interior of an assembly is required by skilled persons for the purpose of adjustment or maintenance, etc., the first objective should be to isolate the assembly from the supply before gaining access. In these situations, a Form 1 'wardrobe' type of assembly (no internal separation) may be suitable unless internal separation is required for other reasons, e.g. to minimise the probability of initiating arc faults. Where isolation is not reasonably practicable for such operations then consideration needs to be given to the specification of an assembly with a higher degree of internal separation. Additional temporary insulating barriers or screens may be required to protect skilled persons from inadvertent contact with energized live parts when working within the assembly. GN1 2.8.1

Electrical assemblies, such as those used for the control of boilers or building services, should always be manufactured to BS EN 60439. Panels which do not comply and are not equipped with suitable internal barriers could allow access by persons who may be electrically unskilled, and may present a serious and totally unnecessary risk of inadvertent or accidental contact with live parts. Barriers need not impede the work activity and may, for example, be made of transparent material, which overcomes the need (or temptation) in many instances to remove them. To further aid safe working, such as adjustment of controls, small holes (affording protection of at least IP2X) may be suitably located in the barriers to permit the insertion of a test probe, or a slim tool (e.g. an insulated screwdriver). A similar approach can often be very effective in preventing shock risks in test bays, test rooms and other situations where barriers are particularly suitable.

612.4.5 Where a barrier or enclosure is provided on site during erection, appropriate testing of the barrier or enclosure must be carried out to verify compliance with BS 7671. However, the protection afforded by a barrier or enclosure must not be impaired during erection or maintenance of equipment. A hole in an enclosure, which should be plugged when not used, may not only destroy any IP classification which the enclosure is required to satisfy, but might also lead to direct contact with internal live parts and could be particularly hazardous where the enclosure is accessible to children. Damage and danger may also result from the entry of vermin through small unplugged openings.

416.2.3 A barrier or enclosure must be firmly secured in place and have sufficient stability and durability to maintain the required degree of protection and appropriate separation from live parts in the known conditions of normal service taking account of relevant external influences.

416.2.5 If, behind a barrier or in an enclosure, an item of equipment such as a capacitor is installed which may retain a dangerous electrical charge after it has been switched off, a warning label must be provided. However, unintentional contact is not considered dangerous if the voltage resulting from static charge falls below 120 V d.c. in less than 5 s after disconnection from the power supply. Small capacitors such as those used for arc extinction and for delaying the response of relays, etc. are not considered to be dangerous.

Automatic disconnection of supply 4

4.1 The protective provisions

The protective measure of automatic disconnection of supply is used in most electrical Sect 411
installations and consists of a basic protective provision and a fault protective provision 411.1
as below.

Protective measure	Protective provisions	
	Basic protection by	Fault protection by
Automatic disconnection of supply	Insulation of live parts, or barriers or enclosures	Protective earthing, protective equipotential bonding, automatic disconnection

In certain circuits or locations additional protection employing 30 mA residual current 410.3.2
devices (RCDs) and/or supplementary bonding must be provided. 411.3.3

4.2 Definitions

A number of terms used in BS 7671 are of particular importance when considering the requirements for automatic disconnection of supply.

4.2.1 System Part 2

It is essential that the type of system of which the proposed installation will form a part be 312.3.1
determined at the earliest stage. Failure to do this could result in the wrong measure(s) being employed or incorrect protective circuit design criteria being applied, either of which could carry potentially serious consequences for the user of the installation at some time during its life. For example, a TT system will require the installation of an earth electrode and the use of RCDs. See also Chapter 11 of this Guidance Note.

4.2.2 Exposed-conductive-part

Conductive part of equipment which can be touched and which is not normally live, Part 2
but which can become live when basic insulation fails.

The term is hyphenated, to emphasise that the words, taken together, have a special meaning as defined in Part 2 of BS 7671. The term must always be stated in full in order to avoid any misunderstanding which can arise from the omission or alteration of any of the words.

Since, by definition, an exposed-conductive-part may become live when basic insulation fails, the term does not include any touchable metallic parts of **Class II** equipment, as such equipment will have supplementary or reinforced insulation to supplement the basic insulation.

The following are everyday examples of exposed-conductive-parts:

- metallic enclosures of switchgear and controlgear assemblies;
- metallic enclosures of wiring systems (conduit, trunking, tray, metallic cable sheaths, etc.);
- metallic wiring accessory boxes and fascias;
- Part 2 metallic enclosures of **Class I** current-using equipment, luminaires, motors, etc.

4.2.3 Extraneous-conductive-part

Part 2 *A conductive part liable to introduce a potential, generally Earth potential, and not forming part of the electrical installation.*

411.3.1.2 Again, the term is hyphenated. The *extraneous-conductive-parts* listed in BS 7671 include non-electrical metallic installation pipes and ducting, which are most often of a distributed nature. Such services, etc. may therefore introduce earth potential alongside much of an electrical installation and the equipment served by it, making it quite likely that a person could be in simultaneous contact with an exposed-conductive-part made live by an earth fault and a nearby extraneous-conductive-part. Bonding of extraneous-conductive-parts to minimise the voltages that could arise between these and exposed-conductive-parts during an earth fault is described in Chapter 13 of this Guidance Note.

4.2.4 Protective conductor (PE)

Part 2 *A conductor used for some measures of protection against electric shock and intended for connecting together any of the following parts:*

(i) Exposed-conductive-parts
(ii) Extraneous-conductive-parts
(iii) The main earthing terminal
(iv) Earth electrode(s)
(v) The earthed point of the source, or an artificial neutral.

It is essential that the designer correctly identifies any particular protective conductor and its specific function, before proceeding to design measures of protection. The
Part 2 figures in Part 2 of BS 7671 should be helpful, and further explanation of the functions of various protective conductors is contained in later chapters of this Guidance Note.

4.2.5 Fault

Part 2 *A circuit condition in which current flows through an abnormal or unintended path. This may result from an insulation failure or a bridging of insulation. Conventionally the impedance between live conductors or between live conductors and exposed- or extraneous-conductive-parts at the fault position is considered negligible.*

4.3 Automatic disconnection of supply — Sect 411

4.3.1 Introduction

Automatic disconnection of supply is the most widely used protective measure for achieving fault protection. Its purpose is to limit (during a fault) the magnitude and duration of the voltage between the exposed-conductive-parts of a circuit and other exposed-conductive-parts or extraneous-conductive-parts or true Earth so as to prevent danger.

Automatic disconnection of supply is a protective measure in which: 411.1

- **i** basic protection is provided by basic insulation of live parts or by barriers or enclosures, and Sect 416
- **ii** fault protection is provided by: 411.3 to 411.6
  - **a** protective earthing,
  - **b** protective equipotential bonding and
  - **c** automatic disconnection in case of a fault.

Figure 4.1 shows a typical protective conductor arrangement for earthing and bonding.

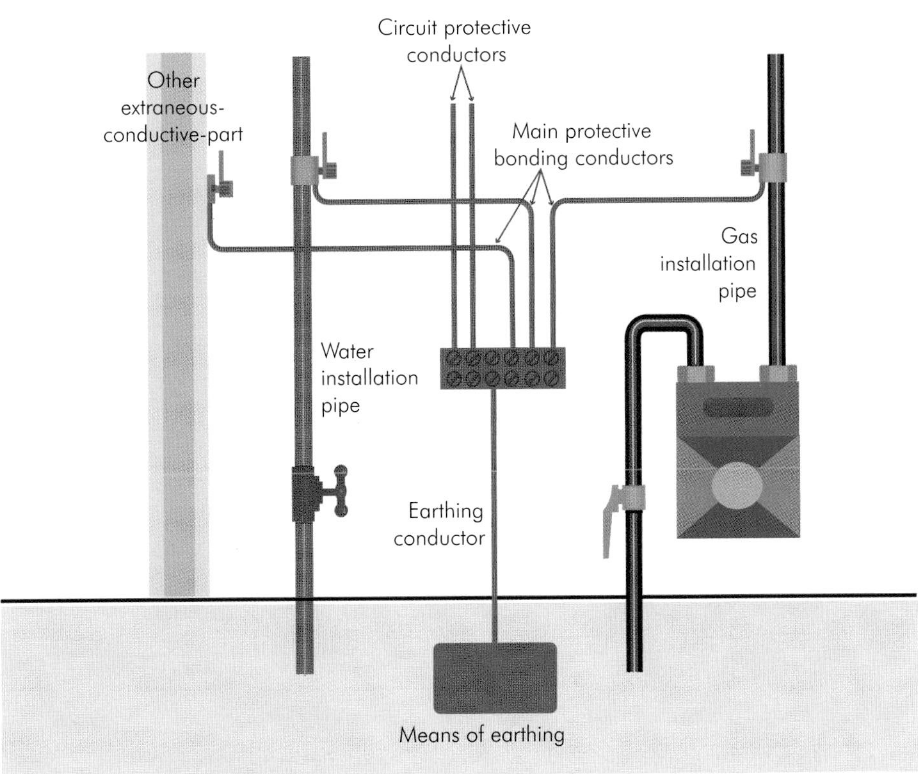

▼ **Figure 4.1** Typical protective conductor arrangement

4.3.2 Protective earthing

Protective earthing of each exposed-conductive-part of the installation, with certain exceptions, is required: Part 2

- **i** to ensure operation of the protective device to disconnect the supply in the event of a fault
- **ii** to limit the rise in potential above Earth potential of exposed-conductive-parts during the fault.

The manner in which the exposed-conductive-parts should be earthed is described, for each type of system, in BS 7671; see also Figures 4.2, 4.3 and 4.4. 411.4.2 411.5.1 411.6.2

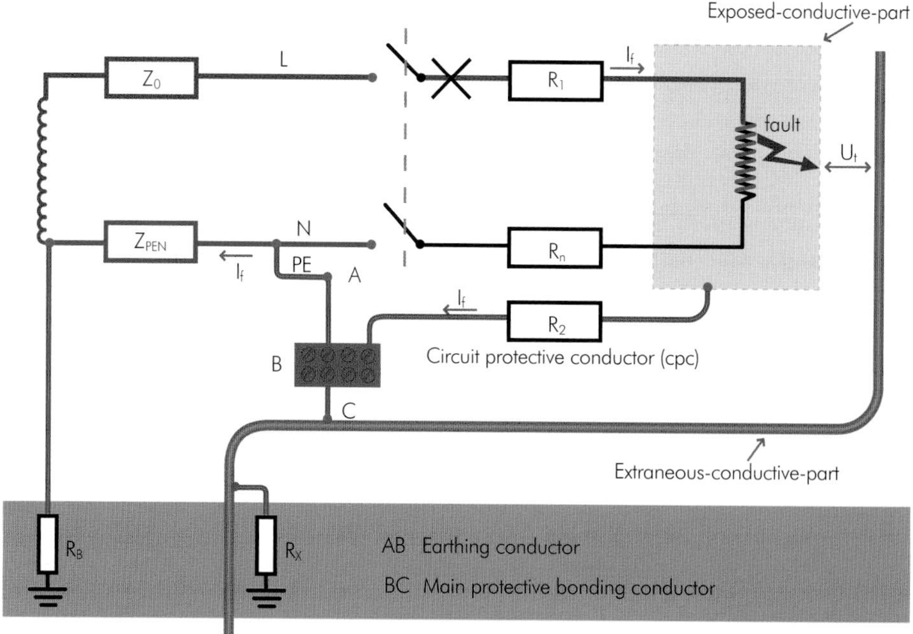

Figure 4.2
Earth fault loop in a TN-C-S system

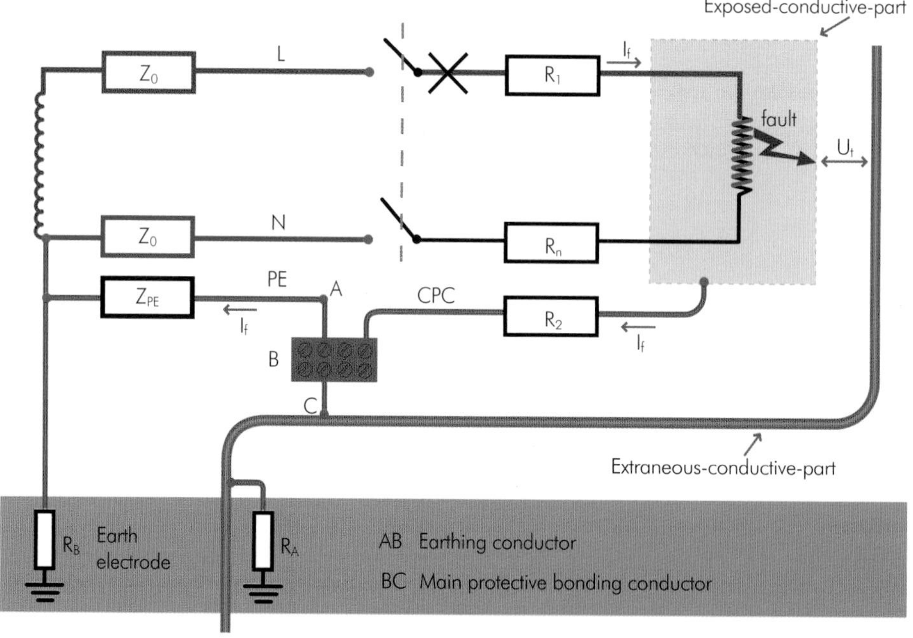

Figure 4.3
Earth fault loop in a TN-S system

The connection with earth referred to above forms part of the *earth fault loop* and, within the installation, it contains two protective conductors: the *circuit protective conductor*
Part 2 *(cpc)* for a particular circuit and the *earthing conductor*; see Figures 4.2 to 4.4. In a TT system (Figure 4.4) the earth return path may not be of particularly low impedance and the protective measures used have to take this into account. Also, especially in a TT system, there may be more than one earthing conductor, each being associated with one or more of the circuits of the installation. The function of both the circuit protective conductor and the earthing conductor is to carry earth fault current, without sustaining damage, until disconnection of the earth fault has taken place. Practical considerations relating to these protective conductors are discussed later in this Guidance Note.

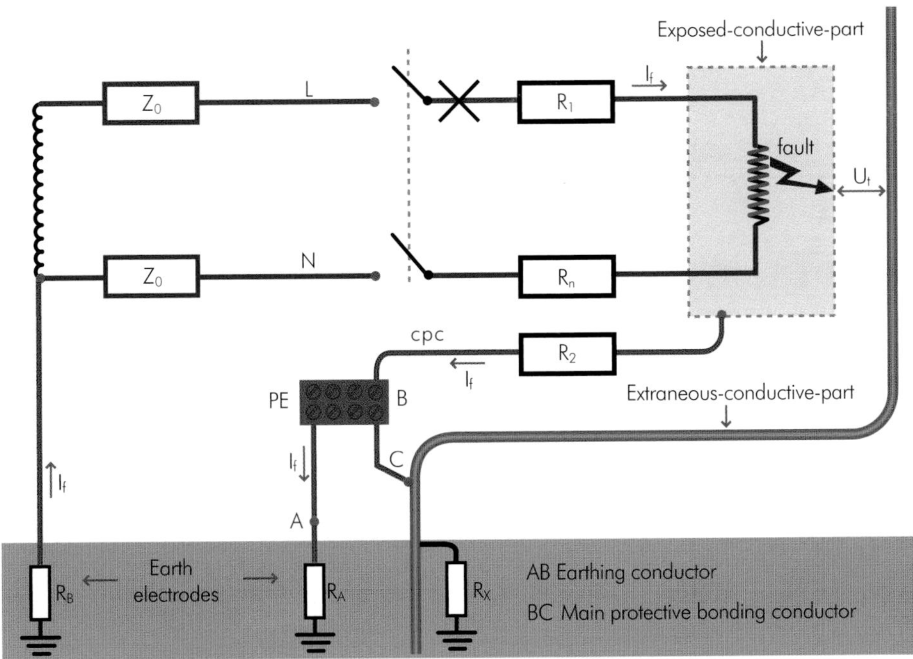

▼ **Figure 4.4** Earth fault loop in a TT system

The importance of *maintaining* good earthing throughout the life of an installation can be most readily appreciated by remembering that failure of the primary protection, i.e. of the basic insulation of a live part, may lead to an exposed-conductive-part presenting a similar danger to that of direct contact with live parts, if the second line of defence is ineffective. GN3

Metal cable support systems, e.g. conduit, trunking, ducting, channelling and trays, require earthing even if they are not used as the protective conductor, unless the live conductors are separated from the support system:

- by basic insulation plus an earthed screen, sheath or armour of current-carrying capacity equal to that of the live conductors, or
- by basic insulation plus supplementary insulation. An example of such supplementary insulation would be plastic conduit but not cable sheathing.

BS 7671 exempts from earthing certain items of metalwork and certain small exposed-conductive-parts. 410.3.9

4.3.3 Protective equipotential bonding

In addition to the earthing of exposed-conductive-parts described above, bonding (connecting together) of these same exposed-conductive-parts and extraneous-conductive-parts is necessary in order to minimise any potential differences that might exist between the parts during an earth fault. The purpose of protective bonding is to equalise potential rather than to carry fault current, although sometimes bonding conductors will also carry fault current where they form part of a parallel earth return path to the source of supply. Part 2

Protective bonding also is effective in the event of an open circuit PEN conductor of a PME supply (or TN-C-S system).

411.3.1.2 Protective equipotential bonding of extraneous-conductive-parts, including:

i water installation pipes
ii gas installation pipes
iii other installation pipework and ducting
iv central heating and air-conditioning systems
v exposed metallic structural parts of the building,

is required in every installation where earthing is employed, yet its importance is often underestimated. To illustrate this, consider the circuit shown in Figure 4.2.

An earth fault in the current-using equipment produces a fault current (I_f) which flows along the circuit protective conductor and back to the source. A small proportion of the current may flow through the main protective bonding conductor directly to Earth, and thence back to the source. The potential difference U_t between the faulty equipment's exposed-conductive-part and the extraneous-conductive-part is:

$U_t = I_f R_2$ (ignoring any reactance of the circuit protective conductor, and any small effect of current flowing in the main protective bonding conductor)

where:

I_f is the fault current
R_2 is the resistance of the circuit protective conductor.

If there is no protective bonding installed, the potential difference U_f between the equipment exposed-conductive-part and the simultaneously accessible extraneous-conductive-part is:

for TN-C-S systems (see Figure 4.2)

$$U_t = I_f (Z_{PEN} + R_2)$$

for TN-S systems (see Figure 4.3)

$$U_t = I_f (Z_{PE} + R_2)$$

for TT systems (see Figure 4.4)

$$U_t = I_f (R_A + R_B + R_2)$$

where:

I_f is the fault current
R_2 is the resistance of the circuit protective conductor
Z_{PEN} is the impedance of the TN-C-S supply PEN conductor
Z_{PE} is the impedance of the TN-S supply PE conductor
R_A is the resistance of the installation earth
R_B is the resistance of the TT supply neutral earth.

For TT systems the touch voltage U_t with no bonding approaches 230 V as R_A and R_B are so large compared with R_1 and Z_0.

4.3.4 Automatic disconnection in case of a fault

When an earth fault occurs, the fault current (I_f) has to be of sufficient magnitude to operate the circuit protective device (fuse, circuit-breaker or RCD) within the required time.

Table 4.1 gives the maximum disconnection times permitted by BS 7671, for nominal line to Earth voltages of 230 V. 411.3.2

▼ **Table 4.1** Maximum disconnection times for a nominal a.c. rms line voltage to Earth, U_0, of 230 V

	Final circuit not exceeding 32 A	Final circuit exceeding 32 A	Distribution circuit	411.3.2 Table 41.1
TN system	0.4 s	5 s	5 s	411.3.2.3
TT system*	0.2 s	1 s	1 s	411.3.2.4

Notes:

* Where, in a TT system, disconnection is achieved by an overcurrent protective device and protective equipotential bonding is connected to all the extraneous-conductive-parts within the installation in accordance with Regulation 411.3.1.2, the maximum disconnection times applicable to a TN system may be used.

- Where disconnection is not required for protection against electric shock it may be required for other reasons, such as protection against thermal effects.
- Where compliance with this regulation is provided by an RCD, the disconnection times in accordance with Table 41.1 relate to prospective residual fault currents significantly higher than the rated residual operating current of the RCD.

A maximum disconnection time of 5 s applies to all circuits in a reduced low voltage system. 411.8.3

A maximum disconnection time of 5 s applies to all circuits supplying fixed equipment used in highway power supplies. 559.10.3.3

4.4 TN system, requirements for disconnection 411.4

In a TN system the characteristics of the protective devices and the circuit impedances are required to fulfil the following requirement: 411.4.5

$$Z_s \times I_a \le U_0$$

where:

Z_s is the impedance in ohms (Ω) of the fault loop comprising:
– the source
– the line conductor up to the point of the fault, and
– the protective conductor between the point of the fault and the source

I_a is the current in amperes (A) causing the automatic operation of the disconnecting device within the time specified in Table 4.1

U_0 is the nominal a.c. rms or d.c. line voltage to Earth in volts (V), usually 230 V.

Maximum values for earth fault loop impedance for a U_0 of 230 V are given in BS 7671:2008 for all the commonly used protective devices where the device standard specifies time/current characteristics, that is:

Table 41.2
Table 41.4
1. Fuses to BS 88, BS 1361, BS 3036 and BS 1362 are given in Table 41.2 for 0.4 s disconnection time and Table 41.4 for 5 s disconnection time

Table 41.3
2. Circuit-breakers to BS EN 60898 and the overcurrent characteristics of RCBOs to BS EN 61009 are given in Table 41.3 for 0.4 and 5 s disconnection times.

Appx 3
Figs 3.4 to 3.6
Note that one set of data only is given for circuit-breakers, because these devices have an operating characteristic with a vertical portion which embraces tripping times in the range 0.1 s to 5 s. The tabulated Z_S values are therefore related to the operating current of that portion of the circuit-breaker characteristic producing tripping in 0.1 s.

The tables of maximum earth fault loop impedances (Z_S) given in BS 7671 are provided for design purposes only (the values need to be reduced for testing purposes). They are based on the worst case disconnection times of circuit-breakers and median times for fuses. Worst case must be taken for circuit-breakers because of their 'dogleg' characteristic. Otherwise other parameters could reduce the fault current to the extent that the disconnection time increased from the instantaneous value 0.1 s to say 10 s
Appx 3 (see Type B circuit-breakers, Figure 3.4 of Appendix 3 of BS 7671). Fuses do not have
Fig 3.4 such characteristics, so small changes in current result in relatively small changes in disconnection time.

The maximum earth fault loop impedances in Tables 41.2 to 41.4 can be calculated from the data in Appendix 3 of BS 7671. For example, consider a 32 A Type B circuit-breaker. Figure 3.4 of BS 7671 gives a current I_a for 0.1 to 5 s disconnection of 160 A.

411.4.5 Using R = V/I, or

$$Z_s = \frac{U_0}{I_a}$$

$$Z_s = \frac{230}{160}$$

$Z_S = 1.4375\ \Omega$, or $1.44\ \Omega$ as tabulated in Table 41.3 of BS 7671.

411.4.4 An RCD may be used to provide fault protection in TN systems where the earth fault
411.4.9 loop impedance (Z_S) is insufficiently low to operate a fuse or circuit-breaker within the prescribed disconnection time. An RCD is also required for additional protection (see section 4.9 of this Guidance Note) and may also be required for installations or
314.1 locations of increased shock risk, such as those in Part 7 of BS 7671. Where more
531.2.9 than one RCD is used they and their circuits should be segregated, or if RCDs are used in series the upstream device needs to be a time-delayed or S type one and the downstream device a type for general use, both to BS EN 61008 or BS EN 61009 to achieve satisfactory discrimination.

4.5 TT system, requirements for disconnection 411.5

Other than in exceptional circumstances in which steel reinforcement of underground concrete or similar structure is available for use, or the ground is perpetually waterlogged, in TT systems it is difficult and costly to achieve a sufficiently low earth electrode resistance for overcurrent devices to be used for fault protection. Regulation 411.5.2 states that RCDs are preferred to overcurrent devices, and RCDs are required for additional protection whichever means is adopted for fault protection, so there are few circumstances where RCDs are not used. 411.5.2 411.3.3

If fault protection is to be provided by an RCD the tabulated values of maximum earth fault loop impedance (Tables 41.2, 41.3 and 41.4) are not applicable. However, overcurrent protective devices are still required to protect the circuits against overload, line to neutral and line-to-line faults.

Where an RCD is used for fault protection, the following conditions are to be fulfilled: 411.5.3

i The disconnection time must be that required by either Table 41.1 or Regulation 411.3.2.4 (Table 4.1 in this chapter), and Table 41.1 411.3.2.4
ii $R_A \times I_{\Delta n} \leq 50$ V

where:

R_A is the sum of the resistances of the earth electrode and the protective conductor connecting it to the exposed-conductive-parts (in ohms)
$I_{\Delta n}$ is the rated residual operating current of the RCD.

Requirements **i** and **ii** above are met if the earth fault loop impedance of the final circuit protected by the RCD meets Table 41.5 of BS 7671:2008, reproduced here as Table 4.2. Table 41.5

▼ **Table 4.2** Maximum earth fault loop impedance (Z_S) to ensure RCD operation in accordance with Regulation 411.5.3 for non-delayed RCDs to BS EN 61008-1 and BS EN 61009-1 for final circuits not exceeding 32 A

Rated residual operating current (mA)	Maximum earth fault loop impedance, Z_s (ohms)†			
	$50\text{ V} < U_0 \leq 120\text{ V}$	$120\text{ V} < U_0 \leq 230\text{ V}$	$230\text{ V} < U_0 \leq 400\text{ V}$	$U_0 > 400\text{ V}$
30	1667*	1667*	1533*	1667*
100	500*	500*	460*	500*
300	167	167	153	167
500	100	100	92	100

† Values for Z_S result from the application of Regulation 411.5.3 (i) and (ii). Disconnection must be ensured within the times stated in Table 41.1 of BS 7671 (Table 4.1 in this chapter).

* The resistance of the installation earth electrode should be as low as practicable. A value exceeding 200 ohms may not be stable. Refer to Regulation 542.2.2.

Requirement **ii** above, $R_A \times I_{\Delta n} \leq 50$ V, does not limit the voltage to 50 V as such, for the fault current I_f must exceed I_a or $I_{\Delta n}$ for the device to operate. However, it does ensure that if the fault current is too low to operate the device, then the voltage between true Earth and the protective conductor will not exceed 50 V.

411.6 4.6 IT systems

▼ **Figure 4.5**
Example of an IT system

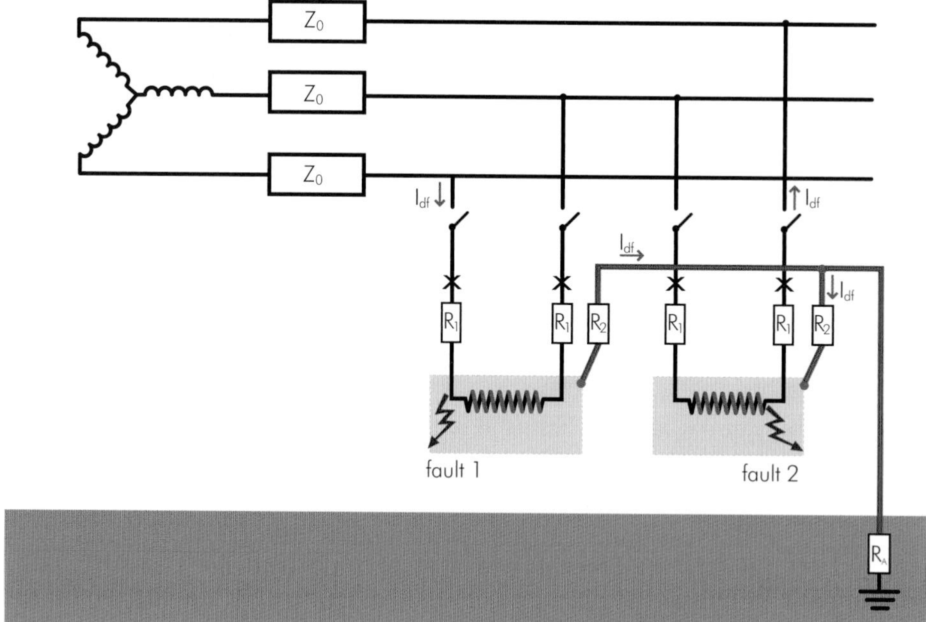

IT systems (see Figure 4.5) are used for electricity distribution in only a few countries, for example Norway. They have the advantage that disconnection is not required on the occurrence of a first fault. While this can provide a reliable supply, such systems suffer from the disadvantage that multiple faults may build up. Disconnection is required on the occurrence of a second fault, but may be difficult to detect. Often residual current monitors/detectors are installed.

▼ **Figure 4.6** Typical IT system with insulation monitoring

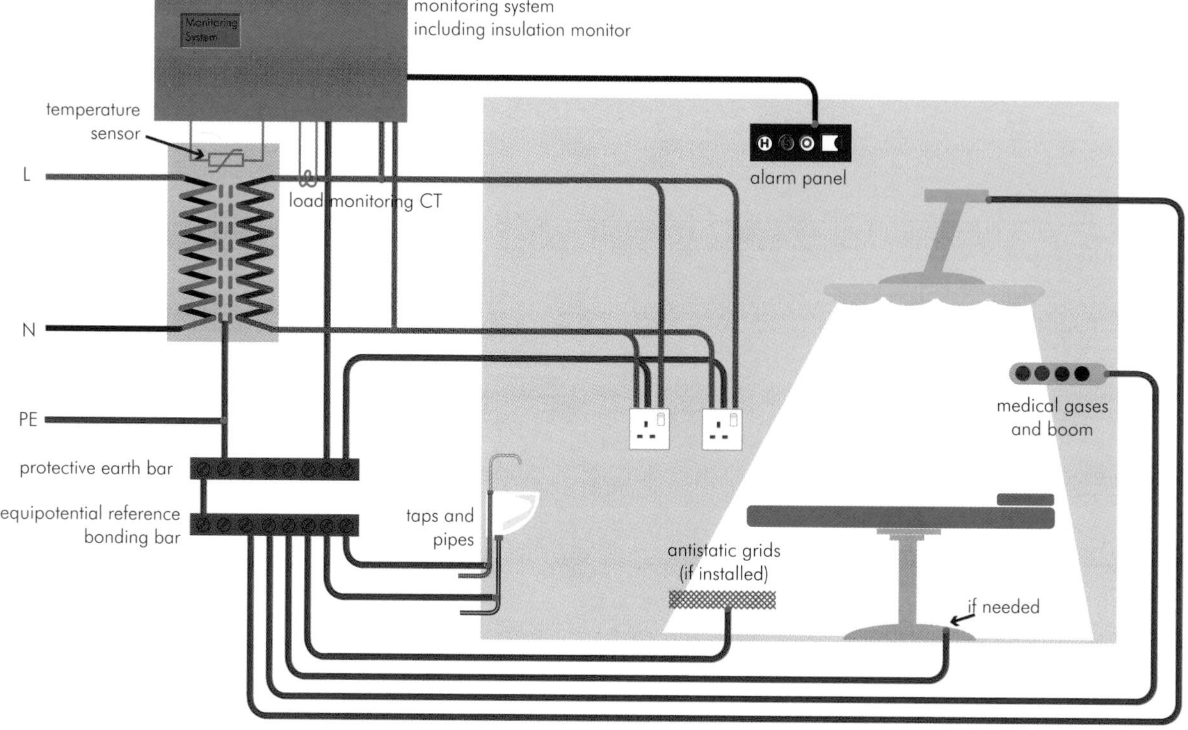

The system is used in specialist installations where reliability of supply is important, as in medical systems for life support; see Figure 4.6. For further information concerning medical locations see Guidance Note 7. GN7

4.7 Functional extra-low voltage (FELV) 411.7

4.7.1 General

Functional extra-low voltage (FELV) installations use the protective measure automatic disconnection of supply of the primary circuit; see Figure 4.7. Part 2

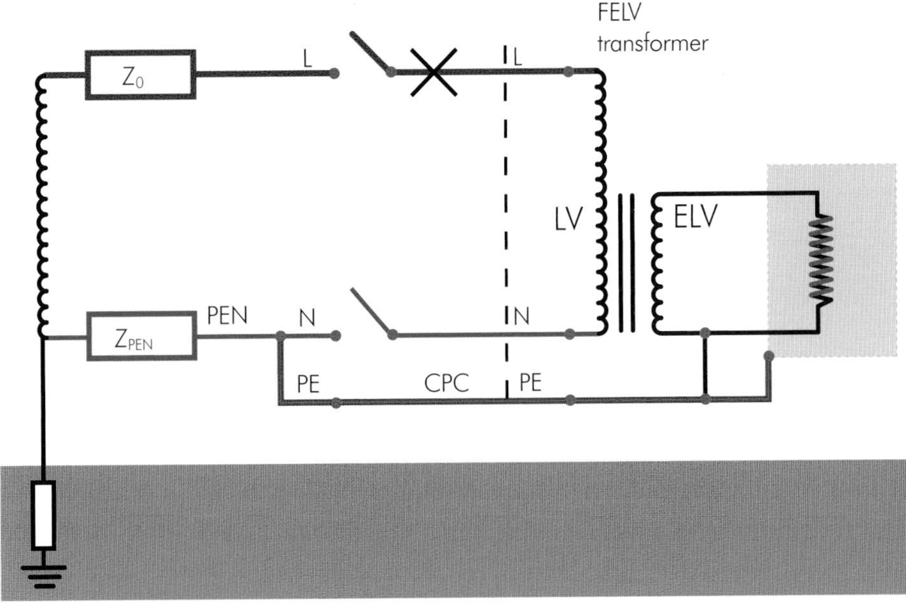

▼ **Figure 4.7** Typical arrangement of LV supply and FELV system

As its name implies, FELV is used when extra-low voltage is required for functional purposes, such as machine control systems.

The supply may be from a transformer with simple separation between the primary and secondary windings, that is, basic insulation. Alternatively, it may be from any of the sources allowed for SELV or PELV, e.g. a safety isolating transformer; see Chapter 7. 411.7.4 414.3

4.7.2 Basic protection

Basic protection for the FELV circuit must be provided by insulation, or barriers or enclosures. 411.7.2

4.7.3 Fault protection

Fault protection is provided by connecting exposed-conductive-parts of the FELV equipment to the protective conductor of the primary circuit, the primary circuit being protected by automatic disconnection of supply.

Fault protection by automatic disconnection of supply in the event of a fault to earth on the FELV system itself is not a requirement. 411.7.3

In practice, most ELV systems require the full range of precautions to be taken against danger arising from any associated low voltage system.

(A method of calculating the loop impedance in a FELV circuit is described in section 7.5.)

411.7.5 **4.7.4 Plugs, socket-outlets, LSCs, DCLs and cable couplers**

Every plug, socket-outlet, luminaire supporting coupler (LSC), device for connecting a luminaire (DCL) and cable coupler in a FELV system must have a protective conductor contact and must not be dimensionally compatible with those used for any other system in use in the same premises.

411.8 4.8 Reduced low voltage systems

4.8.1 Definition

A reduced low voltage system is defined as:

Part 2 *A system in which the nominal line-to-line voltage does not exceed 110 volts and the nominal line to Earth voltage does not exceed 63.5 volts.*

411.8.1.1 The reduced low voltage system described in BS 7671 is the familiar 110 V centre-tap
411.8.1.2 earthed system (see Figure 4.8). The much reduced nominal voltage to Earth (55 V single-phase or 63.5 V three-phase), while being a little above the nominal extra-low voltage range, has provided a high degree of safety for many years on construction sites.

▼ **Figure 4.8**
Typical arrangement of LV supply and reduced low voltage system

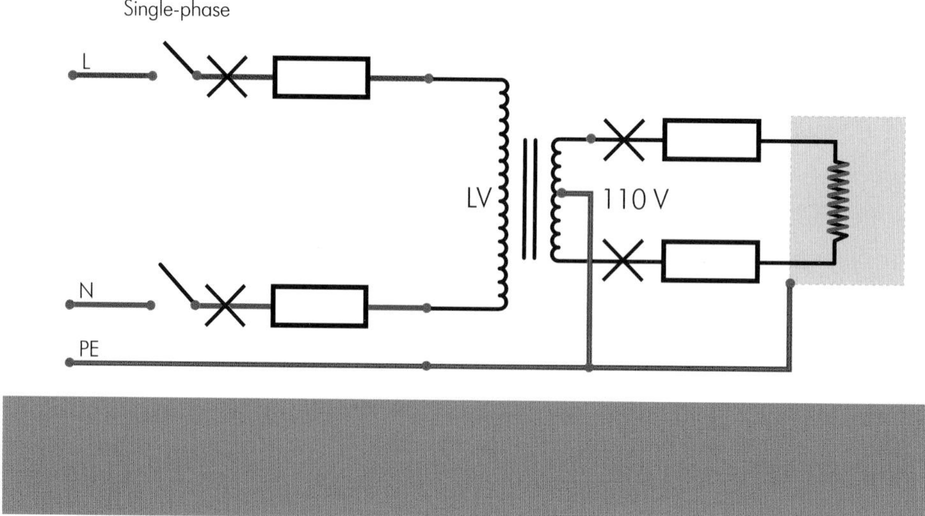

411.8.2 4.8.2 Basic protection

Basic protection must be provided by either basic insulation corresponding to the maximum nominal voltage of the reduced low voltage system or barriers or enclosures.

411.8.3 4.8.3 Fault protection

Fault protection by automatic disconnection of supply must be provided by means of an overcurrent protective device in each line conductor or by an RCD. All exposed-conductive-parts of the reduced low voltage system must be connected to Earth. The earth fault loop impedance at every point of utilisation, including socket-outlets, must be such that disconnection is instantaneous for circuit-breakers and does not exceed 5 s for fuses.

411.4.5 Where a fuse or circuit-breaker is used, the maximum value of earth fault loop impedance (Z_S) may be determined by the following formula:

$$Z_S \times I_a \leq U_0$$

where:

Z_S is the earth fault loop impedance comprising:
– the source
– the line conductor up to the point of the fault, and
– the protective conductor between the point of the fault and the source

I_a is the current in amperes (A) causing instantaneous operation of a circuit-breaker and disconnection within 5 seconds for a fuse. Where an RCD is used this current is the rated residual operating current $I_{\Delta n}$

U_0 is the nominal a.c. rms or d.c. line voltage to Earth in volts (V), that is 55 V for single-phase systems and 63.5 V for three-phase systems.

Alternatively, the values of Z_S specified in Table 41.6 of BS 7671 may be used. Table 41.6

For types and rated currents of devices other than those mentioned in Table 41.6, reference should be made to the appropriate British Standard to determine the value of I_a for the required disconnection time.

4.8.4 Sources

The source of supply to a reduced low voltage circuit must be one of the following: 411.8.4.1

1. A double-wound isolating transformer complying with BS EN 61558-1 and BS EN 61558-2-23
2. A motor-generator having windings providing isolation equivalent to that provided by the windings of an isolating transformer
3. A source independent of other supplies, e.g. an engine driven generator.

The neutral (star) point of the secondary windings of three-phase transformers and generators or the midpoint of the secondary windings of single-phase transformers and generators must be connected to Earth. 411.8.4.2

4.8.5 Circuits

Every plug, socket-outlet, luminaire supporting coupler (LSC), device for connecting a luminaire (DCL) and cable coupler of a reduced low voltage system must have a protective conductor contact and must not be dimensionally compatible with those used for any other system in use in the same premises. Identification of plugs and socket-outlets is usually accomplished by a colour coding system. BS 7375 recommends, for example, that supplies and equipment suitable for an operating voltage between 110 V and 130 V should be coded yellow, as illustrated in Figure 4.9. 411.8.5

▼ **Figure 4.9** Transformer for a reduced low voltage system [photograph courtesy of Blakley Electrics Ltd]

415.1 4.9 Additional protection by residual current devices

(See also Chapter 8.)

415.1.1 415.1.2 BS 7671 recognises this measure as reducing the risk of electric shock in the event of failure of any of the basic protective measures mentioned above (insulation or barriers or enclosures) and/or failure of the provision for fault protection or carelessness by users. The measure must not be used as the sole means of protection.

410.3.2 522.6 Sect 701 Additional protection by means of RCDs is specified as part of a protective measure under certain conditions of external influence, such as concealed cables in walls and partitions, and in certain special locations covered by Part 7, such as circuits supplying equipment in bathrooms.

415.1.1 RCDs for additional protection are required to have a rated residual operating current not exceeding 30 mA and an operating time not exceeding 40 ms at a residual current of 5 $I_{\Delta n}$.

411.3.3 In a.c. systems, additional protection by means of an RCD in accordance with Regulation 415.1, i.e. having an $I_{\Delta n}$ not exceeding 30 mA, must be provided for:

- socket-outlets with a rated current not exceeding 20 A that are for use by ordinary persons and are intended for general use, and
- mobile equipment with a current rating not exceeding 32 A for use outdoors.

An exception is permitted for:

- socket-outlets for use under the supervision of skilled or instructed persons, e.g. in some commercial or industrial locations, or
- a specific labelled or otherwise suitably identified socket-outlet provided for connection of a particular item of equipment. An example would be a single labelled socket-outlet installed to supply a freezer.

415.2 4.10 Additional protection by supplementary equipotential bonding

(See also Chapter 8.)

411.3.2.6 Where automatic disconnection in the time required by Regulation 411.3.2.1 of BS 7671 (4.3.4 in this chapter) cannot be achieved, supplementary equipotential bonding must be provided.

Further information on the applicable requirements for supplementary bonding are covered in section 8.3 of this Guidance Note.

4.11 Where automatic disconnection is not required for shock protection

411.3.2.5 For a system with a nominal voltage U_0 greater than 50 V a.c. or 120 V d.c., automatic disconnection in the time required by Regulation 411.3.2.1 of BS 7671 (4.3.4 in this chapter) is not required if, in the event of a fault to a protective conductor or Earth, the output voltage of the source is reduced in not more than that time to 50 V a.c. or 120 V d.c. or less.

In such a case, consideration must be given to disconnection as required for reasons other than electric shock, for example, overcurrent effects including heating.

4.12 Highway power supplies 559.10.3.3

The maximum disconnection time of 5 seconds allowed for all TN and TT circuits feeding fixed equipment used in highway power supplies may seem at odds with Regulation 411.3.2, requiring say 0.4 second for a 230 V TN final circuit. However, the circuits supplying fixed equipment in highway systems are, in effect, distribution circuits. The final circuit is that supplying the luminaire of a street light from the cut-out, in its column. Even so, there is an anomaly in that Regulation 411.3.2.3 requires a 5 second disconnection time for TN distribution circuits while Regulation 411.3.2.4 requires a 1 second disconnection time for TT distribution circuits.

Double or reinforced insulation 5

5.1 The protective provisions

Sect 412

Protective measure	Protective provisions	
	Basic protection by	Fault protection by
Double insulation	Basic insulation	Supplementary insulation
Reinforced insulation	Reinforced insulation	

412.1.1

This measure requires equipment of particular construction: only equipment with double or reinforced insulation (Class II equipment) or equipment of equivalent construction (declared as equivalent in the product standard). 412.2.1.1

It permits only the use of equipment which has no exposed-conductive-parts that can be made live in the event of a fault and thus cause danger. 412.2.2.1

The measure is generally applicable to individual items of equipment such as switchgear and controlgear assemblies, luminaires and current-using equipment, and to domestic appliances. 412.1.2

5.2 Effective supervision of installations and circuits

The protective measure of double or reinforced insulation may only be used as the sole means of protection against electric shock for an installation or circuit under effective supervision in normal use. This is to ensure that no changes are made that would impair the effectiveness of the protective measure. 412.1.3

In this case the whole installation or circuit should consist entirely of equipment with double insulation or reinforced insulation.

This protective measure must not therefore be applied to any circuit that includes a socket-outlet, luminaire supporting coupler (LSC), device for connecting a luminaire (DCL) or cable coupler, or where a user may inadvertently or in ignorance replace Class II equipment with Class I (requiring an earth).

5.3 Circuit protective conductors

An installation using the protective measure double or reinforced insulation (under effective supervision) does not require a protective conductor and no conductive part shall be connected to a protective conductor.

Otherwise, even when Class II or equivalent equipment is installed, the installation must be provided with circuit protective conductors run to each connection point and accessory. This allows the replacement of Class II equipment with Class I. 412.2.3.2

5.4 Equipment marking

412.2.1.1 Electrical equipment should be type-tested and must be marked with the Class II symbol shown below:

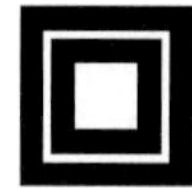

412.2.1.2 Electrical equipment having basic insulation only must have supplementary insulation applied in the process of erecting the electrical installation, providing a degree of safety equivalent to Class II electrical equipment and then marked with the symbol:

412.2.1.3 Electrical equipment having uninsulated live parts must have basic and supplementary insulation applied to it. Where constructional features prevent the application of basic plus supplementary insulation then reinforced insulation may be applied (providing a degree of safety equivalent to Class II equipment). Then the symbol below must be applied:

5.5 Wiring systems

412.2.4.1 Wiring systems (including cables) must have a rated voltage not less than the nominal voltage of the system and at least 300/500 V and be enclosed by:

i a non-metallic sheath, or
ii non-metallic conduit, non-metallic trunking or non-metallic ducting.

Electrical separation 6

6.1 The protective provisions

Sect 413
Sect 418

Protective measure	Protective provisions	
	Basic protection by	Fault protection by
Electrical separation for one item of equipment	Insulation of live parts, or barriers or enclosures	One item of equipment only. Simple separation from other circuits and Earth

Electrical separation is a protective measure in which: 413.1.1

- basic protection is provided by basic insulation or barriers
- fault protection is provided by simple separation of the separated circuit from the primary circuit, other circuits and Earth.

Note: Where there is more than one socket-outlet or one item of equipment, the requirements of Regulation 418.3 apply including supervision by skilled or instructed persons and a prominent warning notice; see sections 6.3 and 10.3 of this Guidance Note. 418.3

In the event of failure of the basic protection, fault protection is provided by the simple separation from the primary circuit so fault current cannot flow to earth and a shock cannot be received. However, the fault is likely to be undetected and in the event of a second fault, a shock risk may arise.

The requirements are different to those in the 16th Edition for 'electrical separation' in that the source must provide only simple separation. If a transformer is used it does not have to be an isolating transformer.

Perhaps the most common use of this measure is with small portable generators, where the winding of the generator is separated from the protective conductor, Earth and the case of the generator. These are often fitted with an RCD, but it is unlikely to operate in the event of a fault.

6.2 Electrical separation applied to one item of equipment

▼ **Figure 6.1**
Small generator supplying one socket-outlet

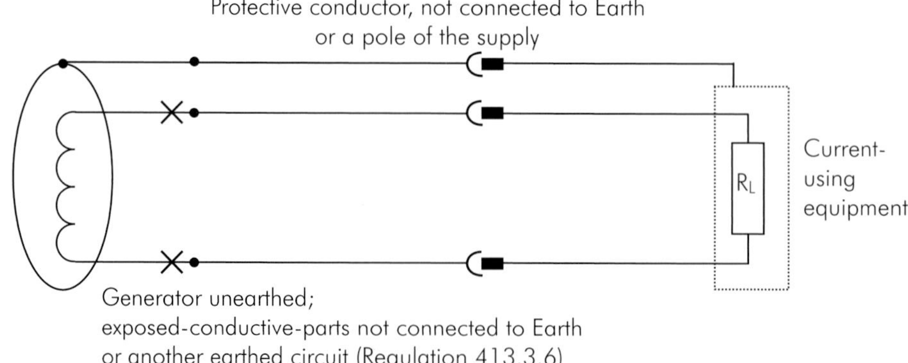

Summary of requirements:

413.3.2 **1** The separated circuit must be supplied through a source with at least simple separation
413.3.2 **2** The voltage of the separated circuit must not exceed 500 V
413.3.3 **3** Live parts of the separated circuit must not be connected at any point to another circuit or to Earth or to a protective conductor
413.3.4 **4** Flexible cables and cords shall be visible throughout any part of their length liable to mechanical damage
413.3.5 **5** The separated circuit should be kept separate from other circuits
413.3.6 **6** No exposed-conductive-part of the separated circuit must be connected either to the protective conductor or exposed-conductive-parts of other circuits, or to Earth.

418.3 6.3 Electrical separation for the supply to more than one item of current-using equipment

▼ **Figure 6.2**
Small generator supplying more than one socket-outlet

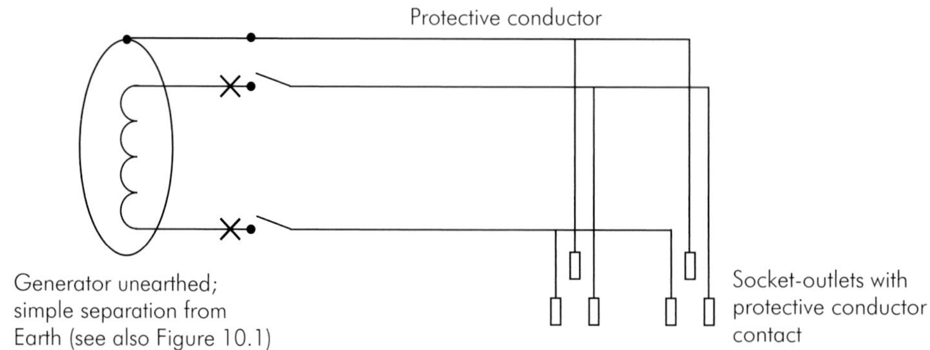

The risk of an undetected fault increases with the number of items of equipment supplied, and the opportunity for fault combinations that are potentially dangerous also increases.

410.3.6 The measure requires the control or supervision of skilled or instructed persons to prevent unsafe changes being made and to check that the installation is properly maintained.

Overcurrent protection is required such that in the event of a fault between the two line conductors (either directly or via a combination of faults), the supply will be disconnected in the time required by Table 41.1, 0.4 second for 230 V supplies. The line-to-line loop impedance must meet the requirements of Tables 41.2 or 41.3. 418.3.7

A warning notice must be fixed in a prominent position adjacent to every point of access to the location concerned. 418.3

514.13.2

> The protective bonding conductors associated with the electrical installation in this location
>
> MUST NOT BE CONNECTED TO EARTH
>
> Equipment having exposed-conductive-parts connected to earth must not be brought into this location.

Extra-low voltage provided by SELV or PELV 7

7.1 The protective provisions

Sect 414

Protective measure	Protective provisions	
	Basic protection by	Fault protection by
Extra-low voltage (SELV or PELV)	Limitation of voltage, protective separation, basic insulation	

Extra-low voltage is defined in BS 7671 as a *nominal voltage not exceeding 50 V a.c. or 120 V ripple-free d.c., whether between conductors or to Earth.* Part 2

Whilst by definition direct current ELV must be 'ripple-free', this term is no longer defined in BS 7671. The meaning was given in Regulation 411-02-09 of the 16th Edition as:

> *For the purposes of this Regulation 'ripple-free' means, for sinusoidal ripple voltage, a ripple content not exceeding 10% rms; the maximum peak value shall not exceed 140 V for a nominal 120 V ripple-free d.c. system and 70 V for a nominal 60 V ripple-free d.c. system.*

This is illustrated in Figure 7.1.

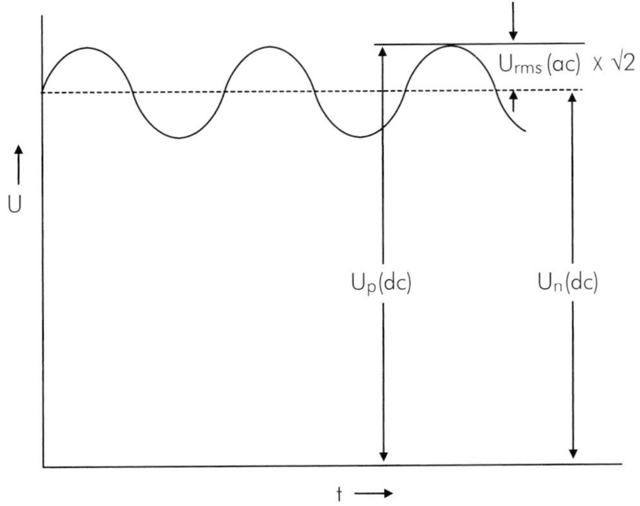

Figure 7.1 Approximate calculation of permissible ripple content of nominal d.c. voltages

Permissible peak value (U_p) is given by:

$$U_p(dc) = U_n(dc) + U_{rms}(ac) \times \sqrt{2}$$

where $U_n(dc)$ = nominal d.c. voltage.

Example for 10 per cent ripple content and U_n(dc) = 120 V:

$$U_p(dc) = 120\ V + 12\ V \times \sqrt{2} = 137\ V, \text{ which approximates to } 140\ V.$$

Excessive ripple can undermine the otherwise relative safety of extra-low voltage (ELV) described below, as it will tend to increase the risk of heart fibrillation (uncoordinated muscular contractions of the heart) arising from an electric shock.

Extra-low voltage may be used either because the equipment or process concerned is designed to operate at ELV, or as a deliberate measure to reduce the danger of electric shock. Regarding the latter, the risk of harm from electric shock at extra-low voltage is very much reduced when compared to a similar situation involving low voltage at, say, 220-240 V to earth. This will be readily appreciated from a study of the information given in IEC publication DD IEC/TS 60479 *Effects of current on human beings and livestock* (see Chapter 2).

411.4.5 The protective measures to be taken in using ELV are dependent upon the location, the source of supply and the degree of separation from other electrical systems. For locations of increased shock risk the particular requirements of Part 7 of BS 7671 must be met. While some aspects of Part 7 are covered in the final chapter of this Guidance Note, it is convenient to consider here, in addition to the general application of SELV and PELV, some of the particular requirements.

7.2 ELV systems

Part 2 Three ELV systems are recognised and defined in BS 7671:

- SELV = separated extra-low voltage
- PELV = protective extra-low voltage
- FELV = functional extra-low voltage.

411.7 FELV is used when extra-low voltage is required for functional purposes, such as machine control systems. The fault protective measure is automatic disconnection of supply of the primary circuit, so FELV is described in section 4.7.

414.3 Regulation 414.3 lists the allowed sources for SELV and PELV systems. In Figure 7.2 the source for SELV and PELV is a safety isolating transformer complying with BS EN 61558-2-6.

▼ **Figure 7.2** Extra-low voltage systems

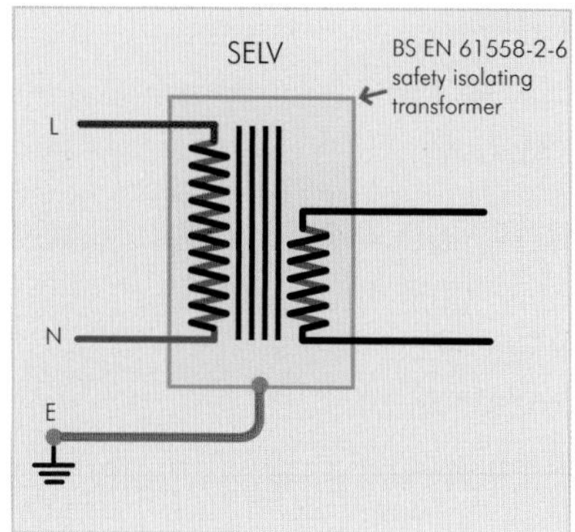

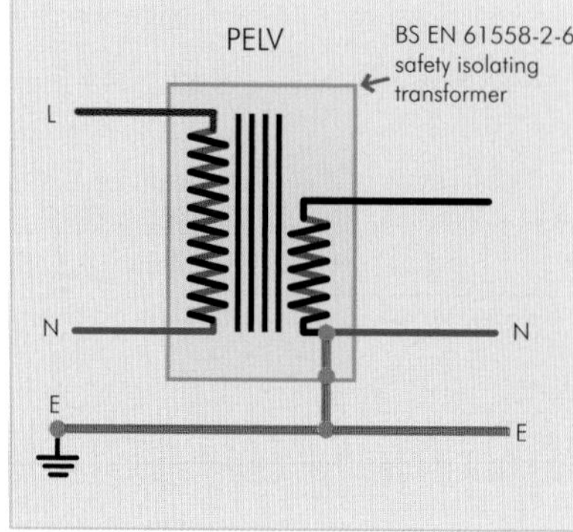

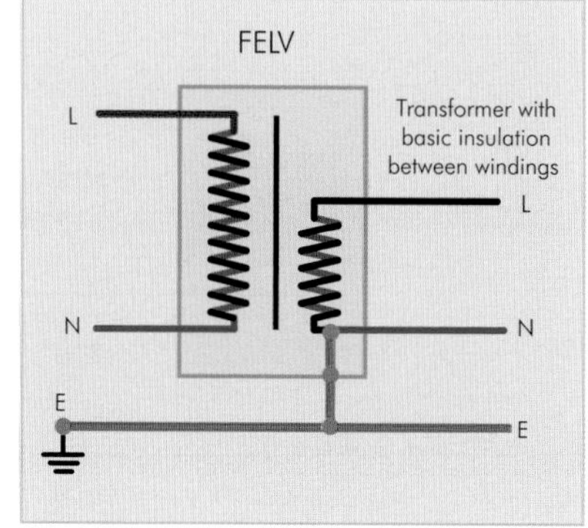

7.3 SELV (separated extra-low voltage)

7.3.1 SELV system

SELV is an ELV system which is electrically separated from Earth and from other 414.1.1
systems in such a way that a single fault cannot give rise to the risk of electric shock. It 414.4.1
should be appreciated that if a single fault were to occur it is intended that this should be confined to the SELV system and not involve any conductive parts of another system. Thus, the construction of a SELV system necessitates the use of high-integrity equipment and materials.

Where an ELV system is derived from a completely independent source, e.g. a battery or an engine-driven generator, and if there is no other electrical system in the vicinity, the precautions to be taken relate only to the ELV system itself.

7.3.2 Protection against contact with live parts 414.4.5

If the nominal voltage of a SELV system exceeds 25 V a.c. or 60 V d.c., basic protection must be provided by insulation of live parts (basic insulation) and, as necessary, the use of barriers or enclosures. At lower voltages, protection is generally afforded by the extra-low voltage itself between live parts.

7.3.3 Application 414.1.2

SELV is most often used where extra-low voltage is specifically chosen as a safety measure, as it does not allow touch voltages from faults elsewhere on the installation to be imported into the location (this is not so with PELV). In certain locations, such as certain zones of swimming pools or certain medical locations, it is the only measure against electric shock permitted and, since even ELV carries some risk of electric shock, the requirements for this measure are also made more stringent, e.g. nominal voltage limited to 12 or 25 V a.c. or 30 or 60 V ripple-free d.c.

7.3.4 The source 414.3

The source for SELV must be of a high standard to minimise the risk of a higher voltage appearing on the conductors of the system. Where the source is a transformer it must be a *safety* isolating transformer complying with BS EN 61558-2-6. There is a difference between a safety isolating transformer and an isolating transformer with simple separation, and the former is specifically intended for SELV circuits.

7.3.5 Separation 414.4.2

Wherever practicable, SELV circuits must be physically separated from those of any other system. If this is impracticable, special precautions are again needed to prevent low voltage appearing on any of the ELV conductors due to a fault between a SELV circuit and another system.

Special insulation resistance tests must also be carried out on completion of the 612.4.1
installation between the SELV and low voltage systems, to verify the integrity of the GN3
precautions taken.

The degree of safety afforded by a SELV system depends upon it being isolated both 414.4.4
from Earth and from any other system. As a result, touch voltages arising from faults elsewhere on the installation cannot be imported into the location by protective conductors.

7.3.6 Overcurrent protection

414.4.3 Since SELV is an unearthed system, overcurrent protective devices must, where required, be fitted in both live conductors. Also, to minimise the risk of danger, socket-outlets in a SELV system are required to be of a pattern which is non-interchangeable with those used for any other system, either low voltage or different ELV, in the same premises, and to have no earth contact (this is to prevent a connection to earth being made in the future, possibly by a person who has little or no knowledge of the system).

7.3.7 Special locations

In certain locations subject to particular requirements, live parts must be insulated and, as appropriate, barriers or enclosures provided, whatever the nominal voltage. Even where precautions against electric shock are not specifically required, insulation of live parts may still be necessary, often with added mechanical protection, to guard against the risk of short-circuits. This is particularly important with the terminals of high-power batteries and where the ELV system involves high-current circuits and hence high energy levels, e.g. electroplating.

7.4 PELV (protective extra-low voltage)

7.4.1 PELV system

414.1.1 PELV is an ELV system which is not electrically separated from Earth, but otherwise
414.4.1 satisfies all the other requirements for SELV; see Figure 7.2.

414.4.5 7.4.2 Protection against contact with live parts

If the nominal voltage of a PELV system exceeds 25 V a.c. or 60 V d.c., protection against contact with live parts must be provided by insulation of live parts and, as necessary, the use of barriers or enclosures.

In normally dry locations protection against contact with live parts may not be required if:

i the voltage does not exceed 25 V a.c. or 60 V d.c. and
ii exposed-conductive-parts are connected by a protective conductor to the main earthing terminal.

In other locations protection against contact with live parts may not be required if the voltage does not exceed 12 V a.c. or 30 V d.c.

GN7 In certain special locations including medical locations protection by insulation of live parts or by barriers or enclosures is essential at all times.

7.4.3 Earth faults on the primary circuit

PELV circuits rely for fault protection against faults on the primary circuit, on the primary circuit protection. It must be confirmed that this fault protection is appropriate for the location.

Faults elsewhere on the installation will introduce fault voltages into the PELV system via the protective conductor.

7.5 Calculation of loop impedance in PELV (and FELV) circuits

If it is wished to ascertain the loop impedance by calculation, the impedance at the end of a circuit fed from the secondary of a step-down transformer is given by:

$$Z_{sec} = Z_p \times \left(\frac{V_s}{V_p}\right)^2 + \frac{Z\%tran}{100}\frac{(V_s)^2}{VA} + (R_1 + R_2)_s$$

where:

Z_p is the loop impedance of the primary circuit including that of the source of supply, Z_e
Z% is the percentage impedance of the step-down transformer
VA is the rating of the step-down transformer
V_s is the secondary voltage
V_p is the primary voltage
$(R_1 + R_2)_s$ is the secondary circuit line and protective conductor resistances.

If data on the step-down transformer is not available, this may be simplified to:

$$Z_{sec} = 1.25\left\{Z_p \times \left(\frac{V_s}{V_p}\right)^2 + (R_1 + R_2)_s\right\}$$

$$Z_{sec} = 1.25\left\{(Z_e + (2R_1)_p) \times \left(\frac{V_s}{V_p}\right)^2 + (R_1 + R_2)_s\right\}$$

where:

1.25 is a factor to compensate for an underestimate
Z_e is the external line-neutral/earth loop impedance
$(2R_1)_p$ is the primary circuit line plus neutral conductor impedance.

7.6 SELV and PELV requirements in Part 7 of BS 7671

Table 7.1 lists particular requirements for SELV and PELV in Part 7 of BS 7671:2008. See also Table 15.2.

▼ **Table 7.1** Summary of main SELV and PELV requirements in Part 7

Part 7 Section		Requirements	Regulations
701	Locations containing a bath or shower	Protection against electric shock	701.414
		Switchgear, controlgear and accessories	701.512.3
		Current-using equipment	701.55
		Floor heating systems	701.753
702	Swimming pools and other basins	Protection against electric shock	702.410.3.4 702.414
		Wiring systems – junction boxes	702.522.24
		Switchgear and controlgear	702.53
		Current-using equipment	702.55.1
		Equipment in zone 1	702.55.4
703	Rooms and cabins containing sauna heaters	Protection against electric shock	703.414
704	Construction and demolition site installations	Protection against electric shock	704.410.3.10 704.414
705	Agricultural and horticultural premises	Protection against electric shock	705.414
706	Conducting locations with restricted movement	Protection against electric shock	706.410.3.10 706.414
711	Exhibitions, shows and stands	Protection against electric shock	711.414
712	Solar photovoltaic (PV) power supply systems	Protection against electric shock	712.414
717	Mobile or transportable units	Additional protection (exception from)	717.415
740	Temporary electrical installations for structures, amusement devices and booths at fairgrounds, amusement parks and circuses	Additional protection (exception from)	740.415

Additional protection 8

8.1 The protective provisions

Additional protection may be required by the particular protective measure, e.g. automatic disconnection of supply, conditions of external influence or in certain special locations covered by Part 7. The additional protective provisions are the installation of: Sect 415

i residual current devices (RCDs)
ii supplementary equipotential bonding.

8.2 Additional protection by residual current devices (RCDs)

8.2.1 General

The use of RCDs with a rated residual operating current ($I_{\Delta n}$) not exceeding 30 mA and an operating time not exceeding 40 ms at a residual current of 5 $I_{\Delta n}$ is recognised in a.c. systems as providing additional protection against electric shock in the event of failure of the provision for basic protection and/or the provision for fault protection or carelessness by users. However, the use of such devices is not recognised as a sole means of protection and does not obviate the need to apply one of the main protective measures. 415.1.1 415.1.2

Thus, any of the primary measures described earlier may be supplemented by the use of an RCD to provide additional protection to persons who come into contact with a live conductor and earth, but it is stressed that an RCD offers no protection at all for a person making direct contact simultaneously with two live conductors (e.g. hand to hand) having a difference of potential between them, unless that person is also in contact with earth.

An RCD will not actually prevent shock currents flowing through the body, for a current must flow to earth for the RCD to detect. However, provided the RCD is sufficiently sensitive and functionally tested periodically it should operate sufficiently quickly to prevent injury in most instances. 612.10

8.2.2 Requirement for RCDs

RCDs (or RCBOs) are required:

i where the earth fault loop impedance is too high to provide the required disconnection time, e.g. where the distributor does not provide an earth – TT systems (in this instance the use of an RCD or RCBO is fault protection not additional protection) 411.5.2

411.3.3 **ii** generally for socket-outlet circuits in domestic and similar installations

701.411.3.3 **iii** for circuits within locations containing a bath or shower

iv for circuits supplying mobile equipment for use outdoors by means of a flexible cable

522.6.7 **v** for cables without earthed metal covering installed in walls or partitions at a depth of less than 50 mm and not protected by earthed steel conduit or similar

522.6.8 **vi** for cables without earthed metal covering installed in walls or partitions with metal parts (excluding screws or nails) and not protected by earthed steel conduit or the like.

Note: v and **vi** apply where the installation is not under the supervision of skilled or instructed persons.

RCDs may be omitted for:

a specific labelled sockets such as those for a freezer. However, the circuit cables must not require RCDs as per **v** and **vi** above, that is circuit cables must be enclosed in earthed steel conduit or have an earthed metal sheath or be at a depth of at least 50 mm in a wall or partition without metal parts.

b socket-outlet circuits in industrial and commercial premises where the use of equipment and work on the building fabric and electrical installation are under the supervision of skilled or instructed persons.

8.2.3 Socket-outlets

General use

411.3.3 Additional protection by means of a 30 mA RCD must be provided for socket-outlets with a rated current not exceeding 20 A that are for use by ordinary persons and are intended for general use.

An exception is permitted for a socket-outlet for use under the supervision of skilled or instructed persons, e.g. in some commercial or industrial locations, or a specific labelled or otherwise suitably identified socket-outlet provided for the connection of a particular item of equipment.

Supervision by skilled or instructed persons

Regulation 4 of the Electricity at Work Regulations requires:

> (1) All systems shall at all times be of such construction as to prevent, so far as is reasonably practicable, danger.
> (2) As may be necessary to prevent danger, all systems shall be maintained so as to prevent, so far as is reasonably practicable, such danger.
> (3) Every work activity, including operation, use and maintenance of a system and work near a system, shall be carried out in such a manner as not to give rise, so far as is reasonably practicable, to danger.
> (4) Any equipment provided under these Regulations for the purpose of protecting persons at work on or near electrical equipment shall be suitable for the use for which it is provided, be maintained in a condition suitable for that use, and be properly used.

The *Memorandum of Guidance on the Electricity at Work Regulations 1989* (HSR 25) published by the Health and Safety Executive (HSE) advises that Regulation 4 covers in a general way those aspects of electrical systems and equipment and work on or near these which are fundamental to electrical safety.

Regulation 4(2) is concerned with the need for maintenance. HSE advise that the regular inspection and testing of equipment is an essential part of any preventative maintenance programme and that records of maintenance including test results be kept which will enable the condition of the equipment and the effectiveness of maintenance policies to be monitored.

This guidance has resulted in the adoption of portable appliance testing regularly undertaken by all commercial and industrial premises (generally in accordance with the recommendations of the IET publication *Code of Practice for In-Service Inspection and Testing of Electrical Equipment*).

Where a company or undertaking appoints a responsible person whose duty is to ensure that:

- all staff are instructed only to use authorised equipment,
- all staff are instructed to check equipment before use, and
- all equipment is regularly maintained in accordance with recognised codes of practice, such as that of the IET,

then RCD protection to socket-outlets could be omitted. This approach is perhaps most usefully adopted where unwanted tripping of an RCD could be particularly disruptive to business, such as a circuit supplying IT equipment.

RCD protection is also not required for socket-outlets under the supervision of skilled persons, for example where RCDs could result in unwanted tripping and prevent a work activity being properly carried out.

8.2.4 Mobile equipment outdoors

Additional protection by means of a 30 mA RCD must be provided for mobile equipment with a current rating not exceeding 32 A for use outdoors. 411.3.3

There is a risk of flexes being cut or damaged, exposing live conductors, and persons touching such live parts while in contact with Earth.

8.2.5 Concealed cables in walls and partitions

Domestic and similar premises

A cable concealed in a wall or partition must be at least 50 mm from the surface on either side, or: 522.6.6

- **i** have earthed armouring or earthed metal sheath, or
- **ii** be enclosed in earthed steel conduit or trunking, or
- **iii** be provided with mechanical protection sufficient to prevent penetration of the cable by nails, screws and the like (Note: the requirement to prevent penetration is difficult to meet), or
- **iv** be installed either horizontally within 150 mm of the top of the wall or partition or vertically within 150 mm of the angle formed by two adjoining walls or partitions, or run horizontally or vertically to an accessory or other electrical equipment, as shown in Figure 8.1.

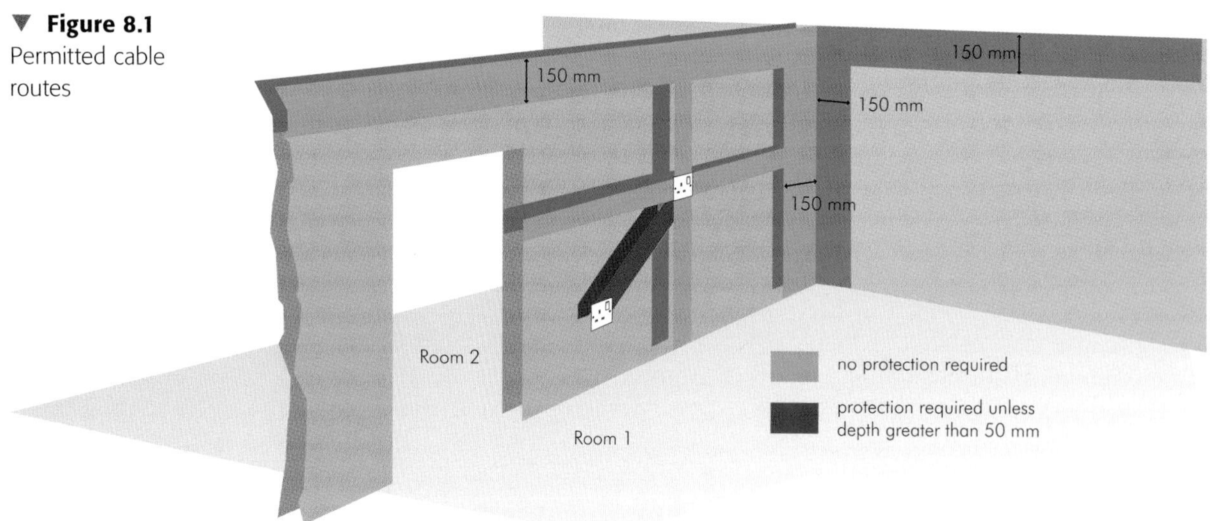

▼ **Figure 8.1**
Permitted cable routes

522.6.7 522.6.8 In domestic and similar installations, cables not installed as per **i**, **ii** or **iii** above must be protected by a 30 mA RCD. Also in such installations any cables, irrespective of their depth, installed in walls or partitions having a metal or part-metal construction must either:

v be installed as **i**, **ii** or **iii** above, or
vi be protected by a 30 mA RCD.

Commercial and industrial premises
For installations under the supervision of a skilled or instructed person, such as commercial and industrial where only authorised equipment is used and only skilled persons will do any work on the building itself, RCD protection of the concealed cables is not required.

As noted in 8.2.3, Regulation 4(3) of the Electricity at Work Regulations requires every work activity, including operation, use and maintenance of a system and work near a system, to be carried out in such a manner as to prevent danger. The guidance from HSE is that this covers work activities of any sort, whether directly or indirectly associated with an electrical system.

It is necessary for an employer to ensure that unskilled persons do not carry out drilling or penetrations of walls or partitions, for example to erect shelves or cabinets or hang pictures. This work must be carried out by persons who are able to avoid all dangers including those associated with cables hidden in walls and partitions.

Designers will need to advise clients that the installation is intended to be under the supervision of skilled or instructed persons. The omission of RCD protection may be an advantage where, for example, unwanted tripping of RCDs could incur cost or danger, e.g. to IT equipment.

8.3 Additional protection by supplementary equipotential bonding 415.2

8.3.1 Application

The notes to Regulation 415.2 advise that supplementary equipotential bonding is considered to be an addition to fault protection. Supplementary bonding may be applied to:

- the whole or part of an installation,
- an item of equipment, or
- a location.

Additional requirements may be necessary for special locations, i.e. as prescribed by Part 7 of BS 7671, or for other reasons.

Supplementary bonding must include all simultaneously accessible exposed-conductive-parts of fixed equipment and extraneous-conductive-parts including, where practicable, the main metallic reinforcement of constructional reinforced concrete. 415.2.1

The equipotential bonding system must be connected to the protective conductors of all equipment including those of socket-outlets.

8.3.2 Automatic disconnection

Where automatic disconnection cannot be achieved in the time required by Regulation 411.3.2.1 then Regulation 411.3.2.6 allows for the use of the circuit arrangement providing supplementary equipotential bonding is carried out. 411.3.2.1 411.3.2.6

Note 2 to Regulation 415.2 advises that the use of supplementary bonding does not exclude the need to disconnect the supply for other reasons, for example protection against fire, thermal stresses in equipment, etc. Designers generally are unhappy with the protection system based on automatic disconnection in case of a fault where this does not result in actual disconnection. The protective provision has a particular use with Type C or D circuit-breakers, where very low loop impedances are required to meet the disconnection times. 415.2 Note 2

8.3.3 Maximum lengths of supplementary bonding conductors

Regulation 415.2.2 requires a further condition to be met: 415.2.2

> **415.2.2** Where doubt exists regarding the effectiveness of supplementary bonding, it shall be confirmed that the resistance R between simultaneously accessible exposed-conductive-parts and extraneous-conductive-parts fulfils the following condition:
>
> $$R \le \frac{50\ \text{V}}{I_a} \text{ in a.c. systems}$$
>
> $$R \le \frac{120\ \text{V}}{I_a} \text{ in d.c. systems}$$
>
> where I_a is the operating current in amperes of the protective device –
>
> for RCDs, $I_{\Delta n}$
> for overcurrent devices, the current causing automatic operation in 5 s.

I_a is not the rating of the protective device (I_n), but the operating current in amperes leading to automatic operation in 5 seconds. For standard protective devices the operating current I_a can be found in Appendix 3 of BS 7671. Figure 3.5 for Type C circuit-breakers from BS 7671 is reproduced here as Figure 8.2. It can be seen that the operating current for a 10 A Type C circuit-breaker is 100 A.

Appx 3
Fig 3.5

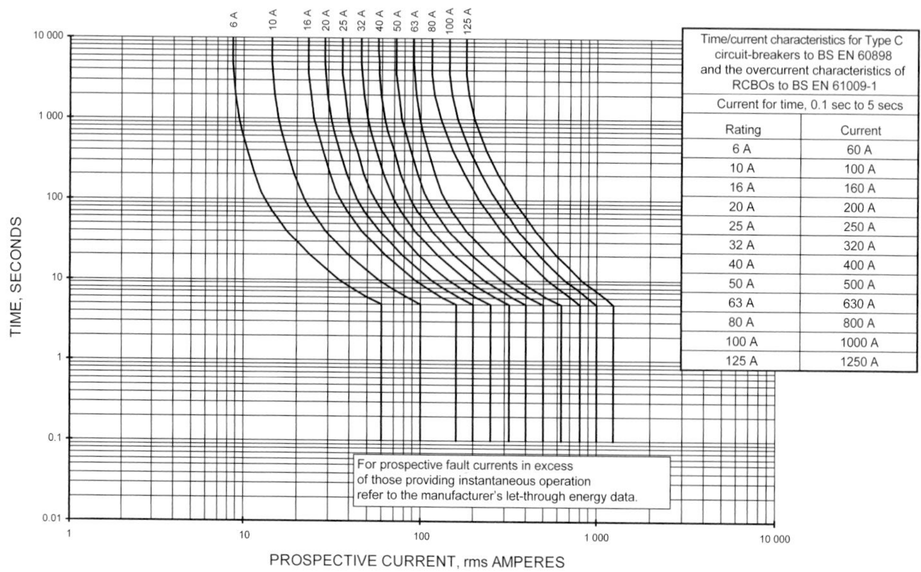

Rating	Current
6 A	60 A
10 A	100 A
16 A	160 A
20 A	200 A
25 A	250 A
32 A	320 A
40 A	400 A
50 A	500 A
63 A	630 A
80 A	800 A
100 A	1000 A
125 A	1250 A

▼ **Figure 8.2**
Operating characteristics for Type C circuit-breakers to BS EN 60898 and RCBOs to BS EN 61009-1 (Figure 3.5 of Appendix 3 of BS 7671)

Example calculation

The maximum length of a, say, 1.0 mm² copper protective conductor (resistance per metre of 18.1 mΩ/m) associated with a 10 A Type C circuit-breaker is calculated below.

I_a is 100 A from table to Figure 8.2.

Using $R \le 50\ \text{V}/I_a$

$$\frac{(\text{Length of cable}) \times 18.1}{1000} \le \frac{50}{100}$$

$$\text{Length of cable} \le \frac{50 \times 1000}{100 \times 18.1}$$

$$\text{Length of cable} \le 27.6\ \text{m}$$

The maximum supplementary bonding conductor resistance R and length L for various supplementary bonding conductor sizes are calculated in Table 8.1.

▼ **Table 8.1** Type C circuit-breakers – maximum length of supplementary bonding conductors to comply with Regulation 415.2.2

Circuit breaker rating, I_n (A)	Current I_a (note 1) (A)	$R = \frac{50}{I_a}$ (Ω)	Area, S_a (note 2) (mm²)	Conductor resistance, R_2 (note 3) (mΩ/m)	Maximum length (L) of conductor $L = R \times \frac{1000}{R_2}$ (m)
6	60	0.83	1.0	18.10	46
10	100	0.5	1.0	18.10	27.6
16	160	0.312	1.0	18.10	17.2
20	200	0.25	1.0	18.10	13.8
25	250	0.20	1.0	18.10	11
30	300	0.16	1.5	12.10	13
32	320	0.156	1.5	12.10	12.8
40	400	0.125	1.5	12.10	10.3
50	500	0.10	1.5	12.10	8.26
63	630	0.079	2.5	7.41	10.7
80	800	0.0625	4.0	4.61	13.55
100	1000	0.05	4.0	4.61	10.8

Notes:

1 From the table within Figure 3.5 of Appendix 3 of BS 7671.

2 From Table 8.6 of *Electrical Installation Design Guide*.

3 From Table F.1 of *Electrical Installation Design Guide*.

Obstacles and placing out of reach 9

(Protective measures to be used only in installations controlled or supervised by skilled persons.)

9.1 Scope of application

Protection by obstacles or placing out of reach is specifically not permitted in the special installations or locations of BS 7671 listed in Table 9.1. Part 7

▼ **Table 9.1** Special installations or locations where protection by obstacles or placing out of reach is not permitted

Part 7 Section		Regulation
701	Locations containing a bath or shower	701.410.3.5
702	Swimming pools and other basins	702.410.3.5
703	Rooms and cabins containing sauna heaters	703.410.3.5
704	Construction and demolition site installations	704.410.3.5
705	Agricultural and horticultural premises	705.410.3.5
706	Conducting locations with restricted movement	706.410.3.5
708	Electrical installations in caravan/camping parks and similar locations	708.410.3.5
709	Marinas and similar locations	709.410.3.5
711	Exhibitions, shows and stands	711.410.3.5
717	Mobile or transportable units	717.417
721	Electrical installations in caravans and motor caravans	721.410.3.5
740	Temporary electrical installations for structures, amusement devices and booths at fairgrounds, amusement parks and circuses*	740.410.3
753	Floor and ceiling heating systems	753.410.3.5

* Placing out of arm's reach is acceptable for electric dodgems at an operating voltage not exceeding extra-low voltage. 740.55.9

9.2 The protective provisions

Protective measure	Protective provisions	
	Basic protection by	Fault protection by
Obstacles	Obstacles	None
Placing out of reach	Placing out of reach	None

As stated in Regulation 417.1, the protective measures of obstacles and placing out of reach provide basic protection only. On their own they provide no fault protective provision. They are for application in installations, with or without fault protection, 417.1 410.3.5

which are controlled or supervised by skilled persons. In other words these protective measures must be used only in installations (locations) where access is restricted to:

i skilled persons, or
ii instructed persons under the supervision of skilled persons.

Part 2 9.3 Definitions

Skilled person: *A person with technical knowledge or sufficient experience to enable him/her to avoid dangers which electricity may create.*

Instructed person: *A person adequately advised or supervised by skilled persons to enable him/her to avoid dangers which electricity may create.*

Competent person: *A person who possesses sufficient technical knowledge, relevant practical skills and experience for the nature of the electrical work undertaken and is able at all times to prevent danger and, where appropriate, injury to him/herself and others.*

9.4 Obstacles

Part 2 Subject to very stringent rules of application, an *obstacle* may be employed as a measure of protection against unintentional (inadvertent) contact only with live parts. By definition, an obstacle (Figure 9.1) can be deliberately circumvented, and therefore
Part 2 it affords a lesser degree of protection than that provided by a *barrier.*

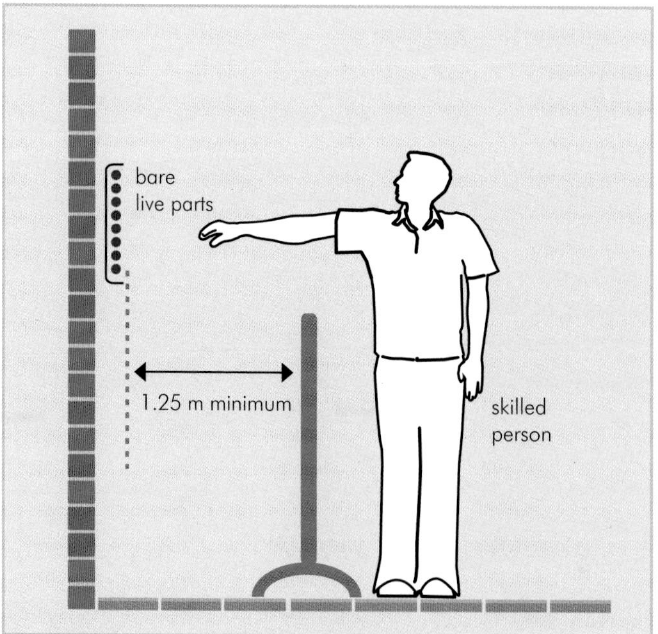

▼ **Figure 9.1**
Obstacle to prevent unintentional bodily approach to or contact with live parts

417.2
410.3.5

The measure is only to be used in locations accessible to skilled persons or instructed persons under the direct supervision of skilled persons. For some installations or locations of increased shock risk this measure is not to be used; see Part 7 of BS 7671.

An obstacle may be fixed or removable and is not intended to prevent 'live working'. Live working falls within the scope of the statutory regulations, in particular Regulation 14 of the Electricity at Work Regulations. The very nature of an obstacle makes it virtually impossible to prescribe a standard test method, and no test is offered in BS 7671.

With this measure of protection there are certain provisos. An obstacle must prevent unintentional bodily approach to live parts and unintentional contact with live parts during the operation of live equipment in normal service. An obstacle must have rigidity and be incapable of unintentional removal. It must also: 417.2.1 417.2.2

1 where removable and as necessary, be preferably constructed of non-conductive material,
2 be of adequate dimensions, and
3 be secured at a sufficient minimum distance from bare live parts (see 9.5.2 Arm's reach). 417.2.1

An obstacle is permitted to be removable without the use of a tool or key. For example, a handrail could be located in sockets that allow it to be lifted out when maintenance work is required, after suitable isolation.

Where this measure is being contemplated, further advice should be sought, e.g. by reference to the Health and Safety Executive (HSE) *Memorandum of Guidance on the Electricity at Work Regulations 1989* (HSR 25). It is worth bearing in mind that, following an accident, the installation designer may be required to justify that protection by obstacles was the most appropriate protective measure to have employed. In such circumstances it is most unlikely to be upheld as appropriate, and no reference is made to this measure in HSE publication HSR 25.

9.5 Placing out of reach

9.5.1 The requirements

Placing bare live conductors out of reach (see Figure 9.2) is recognised by BS 7671 as a means of providing basic protection. Except for overhead lines (installed in accordance with the Electricity Safety, Quality and Continuity Regulations), the measure is only to be used in areas where access is restricted to skilled persons or persons under the supervision of skilled persons. 417.1 410.3.5

Where overhead lines are used for distribution, for example between buildings, they should be installed to the standard of the Electricity Safety, Quality and Continuity Regulations 2002; see 9.5.4.

This limitation is not applied to bare or insulated overhead supply conductors, for which minimum heights above ground and other requirements must be satisfied as laid down in Part V of the Electricity Safety, Quality and Continuity Regulations 2002 (Schedule 2). See also 'Notes on methods of support for cables, conductors and wiring systems' appended to the *On-Site Guide* and Guidance Note 1: *Selection & Erection.*

The minimum height dimension required by BS 7671 for other than an overhead line corresponds to arm's reach, or 2.5 m (see 9.5.2). Fig 417

If the measure is to be used within an installation, considerable thought has to be given to the location of the live parts, in relation to all work activities undertaken in 417.3

▼ Figure 9.2
Placing out of reach to prevent unintentional contact with live parts

417.3

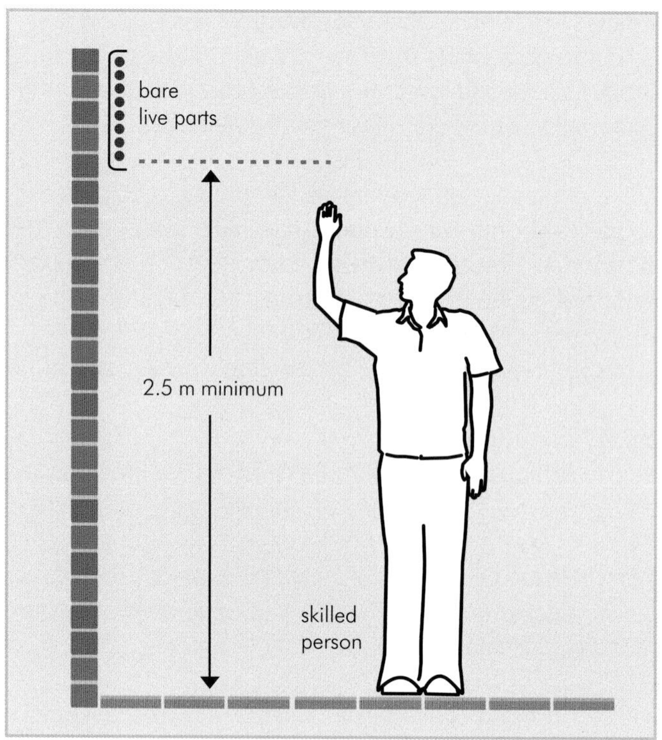

their vicinity. Production work involving the handling of long metal tubes, bars, etc. is likely to present a high degree of risk, making it almost impossible to justify the use of placing out of reach as a protective measure where such work is undertaken. Maintenance activities that may involve the handling of long conducting objects must similarly be carefully considered, whether on the electrical installation itself, the fabric of the building (if any) or on any nearby machines or plant; see Figure 9.3.

▼ Figure 9.3
Increased distances may be required where bulky or long conductive objects are handled

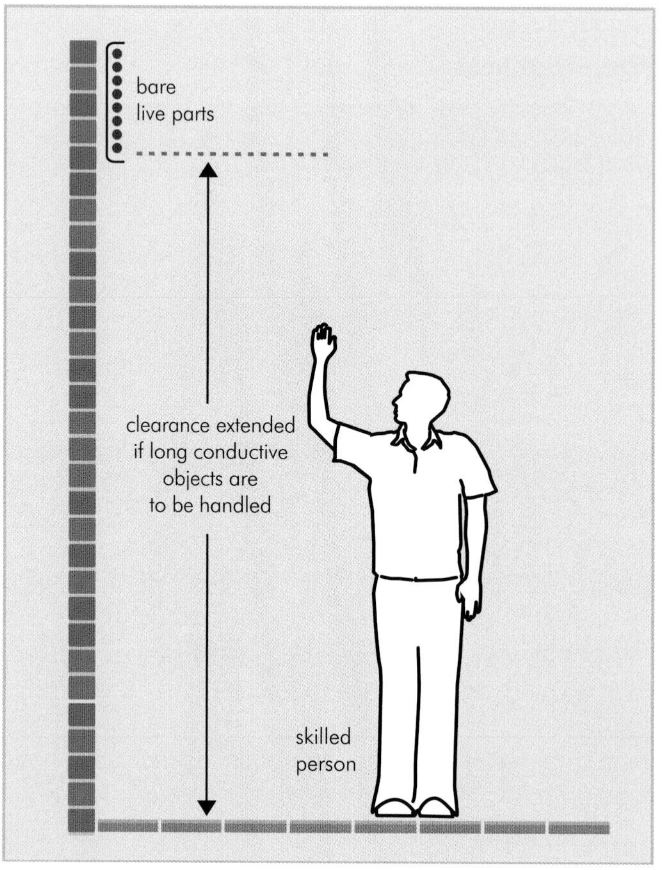

Consideration must also be given to the following:

1 Reduce to a practicable minimum the amount of exposed live material
2 Simultaneously accessible parts at different potentials must not be within arm's reach. Hence adequate minimum distances to all exposed live parts must be ensured 417.3.1
3 A bare live part other than an overhead line must not be within arm's reach or within 2.5 m of an exposed-conductive-part, an extraneous-conductive-part or a bare live part of any other circuit. Two parts are deemed to be simultaneously accessible if they are not more than 2.5 m apart
4 Suitable warning notices should be fixed at all likely points of access to the location
5 Control access to the location, using appropriate means, e.g. locks, written permits
6 Provide any special training for all skilled persons and persons who will be under the supervision of skilled persons who are permitted access to the location; also, provide any protective equipment or clothing that may be necessary during their work
7 Working space, access and lighting must be sufficient such that work can be carried out without risk of danger.

Where protection by placing out of reach is being considered, further advice should be sought where necessary, e.g. by reference to HSE. As discussed in 9.4, regarding the use of obstacles, it is worth bearing in mind that, following an accident, the installation designer may be required to justify that protection by placing out of reach was the most appropriate protective measure to have employed.

9.5.2 Arm's reach

Arm's reach is defined as: Part 2

A zone of accessibility to touch, extending from any point on a surface where persons usually stand or move about to the limits which a person can reach with a hand in any direction without assistance.

Figure 417 of BS 7671 is reproduced here as Figure 9.4. This shows that the limit of arm's reach is normally taken as 1.25 m horizontally and 2.5 m vertically and, as previously mentioned, these dimensions may need to be increased where bulky or long objects are handled, such as metal ladders. Fig 417

If a normally occupied position is restricted in the horizontal direction by an obstacle (e.g. handrail, mesh screen) affording a degree of protection less than IPXXB or IP2X, arm's reach is considered to extend from that obstacle. 417.3.2

In the overhead direction, arm's reach is 2.5 m from the surface S in Figure 9.4, not taking into account any intermediate obstacle providing a degree of protection less than IPXXB. The values of arm's reach apply to contact directly with bare hands without assistance (e.g. tools or ladder).

▼ **Figure 9.4**
Arm's reach

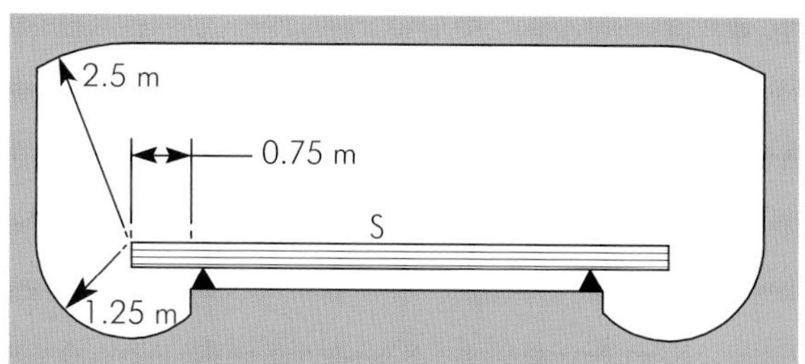

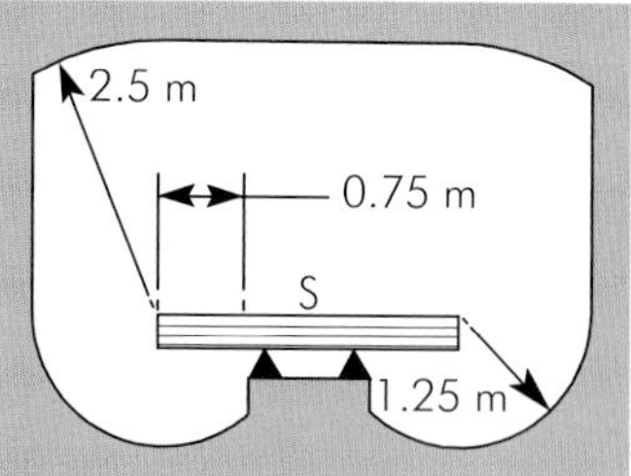

S = surface expected to be occupied by persons

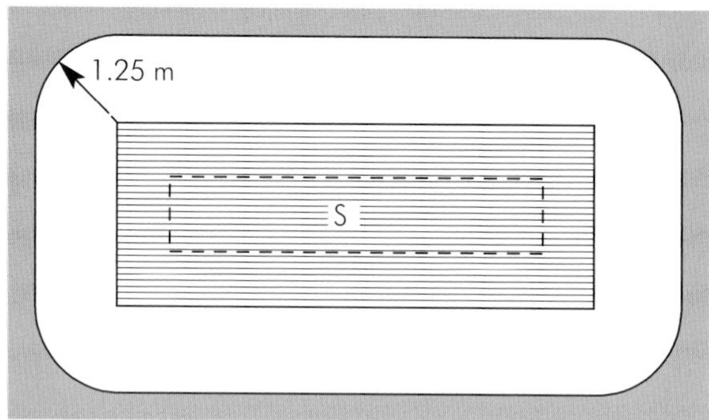

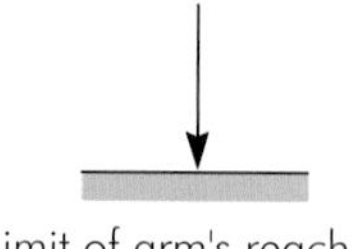

limit of arm's reach

The values refer to bare hands without any assistance, e.g. from tools or a ladder.

9.5.3 Electricity at Work Regulations 1989

Regulation 7(b) has requirements for placing out of reach:

7. Insulation, protection and placing of conductors

All conductors in a system which may give rise to danger shall either –

(a) be suitably covered with insulating material and as necessary protected so as to prevent, so far as is reasonably practicable, danger; or
(b) have such precautions taken in respect of them (including, where appropriate, their being suitably placed) as will prevent, so far as is reasonably practicable, danger.

Guidance is given in the HSE *Memorandum of Guidance on the Electricity at Work Regulations 1989* (HSR 25) as follows:

> *Regulation 7(b) permits the alternative [to (a)] of having other precautions taken in respect of the conductors. These precautions may include the suitable placing of conductors. The precautions may comprise strictly controlled working practices reinforced by measures such as written instructions, training and warning notices etc. The precautions must prevent danger so far as is reasonably practicable. Examples where bare conductors are used in conjunction with suitable precautions are to be found in many applications including overhead electric power lines, down-shop conductors for overhead travelling cranes in factories etc, railway electrification using either separate conductor and running rails or overhead pick-up wires, and certain large electrolytic and electrothermal plants.*

9.5.4 Minimum height of overhead lines, wires and cables

Overhead lines must be installed to the standard required by the Electricity Safety, Quality and Continuity Regulations 2002 (ESQCR) which require, in Regulation 17, that an overhead line, at the maximum likely temperature of that line, shall be at a height not less than the values in Table 9.2. 417.3

▼ **Table 9.2** Minimum heights of overhead lines

Nominal voltage	Over roads	Other locations
Not exceeding 33 000 volts	5.8 metres	5.3 metres

The above requirement does not apply to any section of an overhead line at a point where it is not over a road accessible to vehicular traffic and which:

1 is surrounded by insulation; or
2 is not surrounded by insulation but is at least 4.3 metres above ground and connects equipment mounted on a support to any overhead line; or
3 is connected with earth.

The height above ground of any wire or cable which is attached to a support carrying any overhead line shall not be less than 5.8 metres at any point where it is over a road accessible to vehicular traffic.

Protective measures for application only where the installation is controlled or under the supervision of skilled or instructed persons 10

10.1 Protective measures and provisions

The protective measures specified in Section 418, that is Sect 418 410.3.6

i non-conducting location
ii earth-free local equipotential bonding
iii electrical separation for the supply to more than one item of current-using equipment,

shall be applied only where the installation is under the supervision of skilled or instructed persons so that unauthorised changes cannot be made (see 9.3 for the definitions of skilled and instructed persons).

Protective measures	Protective provisions		
	Basic protection by	Fault protection by	
Non-conducting location	Insulation of live parts, or barriers or enclosures	No protective conductor, insulating floor and walls, spacings/obstacles between exposed-conductive-parts and extraneous-conductive-parts	418.1
Earth-free local equipotential bonding	Insulation of live parts, or barriers or enclosures	Protective bonding, notices, etc.	418.2
Electrical separation for the supply to more than one item of equipment	Insulation of live parts	Simple separation from other circuits and earth, separated protective bonding, etc.	418.3

Part 7 Protection by non-conducting location or earth-free local equipotential bonding is specifically not permitted in the special installations or locations of BS 7671 listed in Table 10.1.

▼ **Table 10.1** Special installations and locations where protection by non-conducting location or earth-free local equipotential bonding is not permitted

Part 7 Section		Regulation
701	Locations containing a bath or shower	701.410.3.6
702	Swimming pools and other basins	702.410.3.6
703	Rooms and cabins containing sauna heaters	703.410.3.6
705	Agricultural and horticultural premises	705.410.3.6
708	Electrical installations in caravan/camping parks and similar locations	708.410.3.6
709	Marinas and similar locations	709.410.3.6
711	Exhibitions, shows and stands	711.410.3.6
712	Solar photovoltaic (PV) power supply systems	712.410.3.6
717	Mobile or transportable units*	717.418
721	Electrical installations in caravans and motor caravans	721.410.3.6
740	Temporary electrical installations for structures, amusement devices and booths at fairgrounds, amusement parks and circuses	740.410.3.6
753	Floor and ceiling heating systems	753.410.3.6

717.418 * For these units, protection by earth-free local equipotential bonding is permitted but not recommended.

418.1 10.2 Non-conducting location

418.1.6 Protection by non-conducting location has been included in BS 7671 since 1981 because it was traditionally used overseas, where it can be supported by the knowledge and understanding of other trades, such as plumbing. Similar support may not generally be available in the UK and accordingly this method of protection is not recognised for general application. If it is intended to employ this measure then the designer is advised to consult the HSE.

This protective measure is an alternative to others in Chapter 41, e.g. Section 411, automatic disconnection of supply. However, the method of protection of non-conducting location permits the use of an item of Class 0 equipment that might have a metal case (exposed-conductive-part) that is not connected to the installation main
418.1.2 earthing terminal. The intent of protection by non-conducting location is to prevent simultaneous contact with conductive parts that may be at different potentials due to insulation failure. A protective device will not operate in the event of a fault to such an exposed-conductive-part.

This protective measure is not to be confused with additional precautions taken in electrical repair shops, test bays, etc. to restrict the accessibility of exposed- or extraneous-conductive-parts. Such locations will be protected primarily by automatic disconnection of supply or electrical separation.

The following notes are provided to assist designers who may be involved in installing non-conducting locations overseas. They outline the principles upon which the method is based, but in no way do they provide a full design guide.

All electrical equipment installed in a non-conducting location must be provided with basic protection. 418.1.1

This measure requires that there is adequate physical separation between exposed-conductive-parts and extraneous-conductive-parts which could be at different potentials if there was a failure of the basic insulation of the equipment concerned. Additionally, fault protection is obtained by the insulating nature of the floor and walls of the location itself. The minimum insulation resistance to be achieved is such that any shock current would be well under 10 mA. To prevent any reduction in the insulation resistance of the walls and especially the floor, the surfaces must be kept clean and dry, and it may be necessary to control the air humidity in the location. 418.1.2 418.1.5 612.5

There must be no protective conductor in a non-conducting location. 418.1.3

The separation distance between conductive parts (whether they are exposed or extraneous) between which a voltage can exist in the event of insulation failure in the equipment concerned, or elsewhere in the system, must be not less than 2.5 m. If the parts are outside the zone of arm's reach the minimum separation between them becomes 1.25 m. Two other arrangements are also acceptable: 418.1.4

1. Effective obstacles, as far as possible made of insulating material, fixed between exposed-conductive-parts and extraneous-conductive-parts
2. Insulation applied to or enclosing extraneous-conductive-parts to prevent access, or inserted to prevent the introduction of a potential (usually earth) into the location. Such insulation would need to be tested. 612.5.2

The continued effectiveness of this measure can be very difficult to guarantee, since there is always the risk that other conductive parts could be introduced at a later date, such as mobile or portable Class I equipment, or metallic water or other service pipes, which would invalidate compliance. There could also be the possibility that a fault voltage appearing on conductive parts within the non-conducting location might be transferred beyond the location via an extraneous-conductive-part, such as a water pipe, which has been added after completion of the installation. Hence the need for implementation in the workplace of effective supervision of the whole location, including all of its electrical equipment and all other equipment and services. For non-conducting locations, BS 7671 does not prescribe any specific notice requirements. Nonetheless, a suitable notice similar to the example below should be fixed in a prominent position adjacent to every point of access to the location. 418.1.6 418.1.7

Caution
Non-conducting location

Person in charge.....................
Only this person may authorise any equipment to be taken into or any work to be undertaken in this location.

418.2 10.3 Earth-free local equipotential bonding

The application of this measure is strictly limited to special situations which are earth-free, and where its use is fully justified on operational grounds. An example would be a medical, special electronic or communications equipment application, where vital functional equipment, while having exposed-conductive-parts (i.e. not being Class II equipment or having equivalent insulation), will not operate satisfactorily if
418.2.2 connected to a means of earthing. The location where the measure is applied may have a non-conducting floor, or a conductive floor which is insulated from Earth, and must have every simultaneously accessible exposed-conductive-part and extraneous-conductive-part within the location connected by local protective bonding conductors.
410.3.6 The measure requires effective supervision and regular inspection and testing of the protective features employed.

418.2.3 **Note:** There must be no extraneous-conductive-parts introducing earth potential into the location.

The correct implementation of the requirements results in a 'Faraday cage' and prevents the appearance of any dangerous potential differences between simultaneously accessible parts within the location concerned. Thus, in this application of equipotential bonding, the use of the word 'equipotential' really does reflect what it means for a first fault on the equipment, although this may not hold true for two simultaneous faults.
418.2.3 The local protective bonding conductors must not be in electrical contact with Earth directly, nor in contact with Earth through exposed-conductive-parts or extraneous-conductive-parts.

This measure cannot readily be applied to an entire installation and it is difficult to coordinate safely with other protective measures used elsewhere in the installation.
418.2.4 In particular, precautions are necessary at the threshold of the earth-free equipotential
418.2.5 location. Also, a warning notice complying with Regulation 514.13.2 must be fixed at every point of access into the location to warn against the importation of an earth (see below).

514.13.2

> The protective bonding conductors associated with the electrical installation in this location
>
> MUST NOT BE CONNECTED TO EARTH.
>
> Equipment having exposed-conductive-parts connected to earth must not be brought into this location.

418.2.1 All equipment used in the location must have basic protection.

The form of supply to the equipment in the location necessitates special consideration. Employing the mains supply of a TN system would import an earth into the location via the earthed neutral conductor which, in the event of a neutral fault to an exposed-conductive-part, would earth the equipment. Earth-free local equipotential bonding is normally associated with electrical separation, which overcomes this problem, but, where two measures of protection are to be used in the same location, care must be taken to ensure that the particular requirements for each measure are fully satisfied and, most importantly, are mutually compatible.

10.4 Electrical separation for the supply to more than one item of current-using equipment 418.3

(See also section 6.3.)

Provided the applicable requirements in BS 7671:2008 are met, it is permitted for a separated circuit to supply two or more items of current-using equipment.

The risk that can arise with protection by electrical separation under such conditions
is that whilst with a first fault there is normally no risk of electric shock, the fault is
likely to be undetected and a hazard may arise upon a second fault or third fault. The 410.3.6
measure therefore requires effective supervision and regular inspection and testing of
the protective features employed.

All electrical equipment supplied by the separated circuit must have basic protection, 418.3.1
such as insulation or barriers and enclosures.

Where a single source supplies several items of equipment, all the requirements
detailed for electrical separation to a single item of current-using equipment given in
Section 413 have to be met and, in addition, the requirements detailed in Regulation 418.3.2
418.3 for electrical separation for the supply to more than one item of current-using
equipment also have to be met.

The separated circuit must be protected from damage and insulation failure. 418.3.3

The separated circuit requires careful design. Overcurrent protection should be provided 418.3.7
in each pole of the separated circuit to provide not only overload and/or short-circuit
protection, but also automatic disconnection of the supply in the event that two faults to
different exposed-conductive-parts occur and these are fed by conductors of different
polarity as illustrated by Figure 10.1. Such a two-fault occurrence would, in effect, be
a short-circuit via the protective bonding conductor(s) connecting together the two
relevant exposed-conductive-parts. If this bonding were missing or not effective, then
the two-fault occurrence could lead to the full supply voltage appearing between the
exposed-conductive-parts, thereby creating a very serious shock hazard to a person
simultaneously in contact with those parts. The protective device must operate within
the time stated in Table 41.1 of BS 7671.

L

N

PE

non-earthed protective bonding conductors

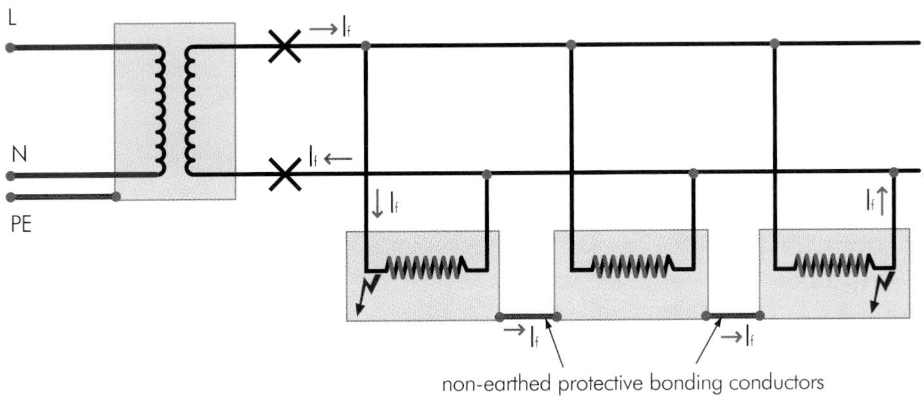

▼ Figure 10.1 Electrical separation to more than one item of equipment

418.3.4 All the exposed-conductive-parts of the separated circuit must be connected together by insulated, non-earthed protective bonding conductors which must not be connected to any of the following:

- the protective conductor of any other circuit
- any exposed-conductive-parts of any other circuit, or
- any extraneous-conductive-parts.

The cross-sectional area of the non-earthed protective bonding conductors must be calculated or selected in order to meet the requirements of Section 543.

418.3.6 All flexible cables must include a protective conductor for use as a protective bonding conductor unless supplying equipment with double or reinforced insulation.

418.3.5 Every socket-outlet must be provided with a protective conductor contact which must be connected to the equipotential bonding system.

418.3.8 The product of the nominal voltage of the circuit in volts and length in metres of the wiring system should not exceed 100000 Vm and the length of the wiring system should not exceed 500 m.

418.3 A warning notice complying with Regulation 514.13.2 must be fixed in a prominent position adjacent to every point of access into the location, to warn against introducing a connection with Earth (see below).

514.13.2

> The protective bonding conductors associated with the electrical installation in this location
>
> MUST NOT BE CONNECTED TO EARTH.
>
> Equipment having exposed-conductive-parts connected to earth must not be brought into this location.

Earthing 11

11.1 Earthing systems Sect 542

11.1.1 Connections to Earth

It is necessary to determine the type of system, i.e. TN-S, TN-C-S, TT, etc., of which the proposed installation will form a part, before proceeding with the installation design. Part 2 312.3.1

System descriptions are given in Part 2 of BS 7671, while Appendix 9 provides descriptions of multiple-source, d.c. and other systems. Appx 9

A distributor is required on request to provide a statement on the type of earthing (Regulation 28 of the Electricity Safety, Quality and Continuity Regulations 2002 (ESQCR)), and, unless inappropriate for reasons of safety, the distributor is required to make available his supply neutral conductor or protective conductor for connection to the consumer's earth terminal (Regulation 24). For a low voltage supply given in accordance with the ESQCR the supply system will be TN-S, TN-C-S (PME) or TT, and most commonly for new supplies TN-C-S. TN-C and IT systems are both very uncommon in the UK, the former because it requires an exemption from the ESQCR which invokes special installation arrangements and the latter because the source is not directly earthed and this is not permitted for a low voltage public supply in the UK (Regulation 8).

For TN-S, TN-C-S and TT systems the following explanations should aid a full understanding of the earthing arrangements and their scope of application. The nomenclature of these system types is as follows:

T = Earth (from the French word *Terre*)
N = Neutral
S = Separate
C = Combined

▼ **Figure 11.1**
System earthing arrangements

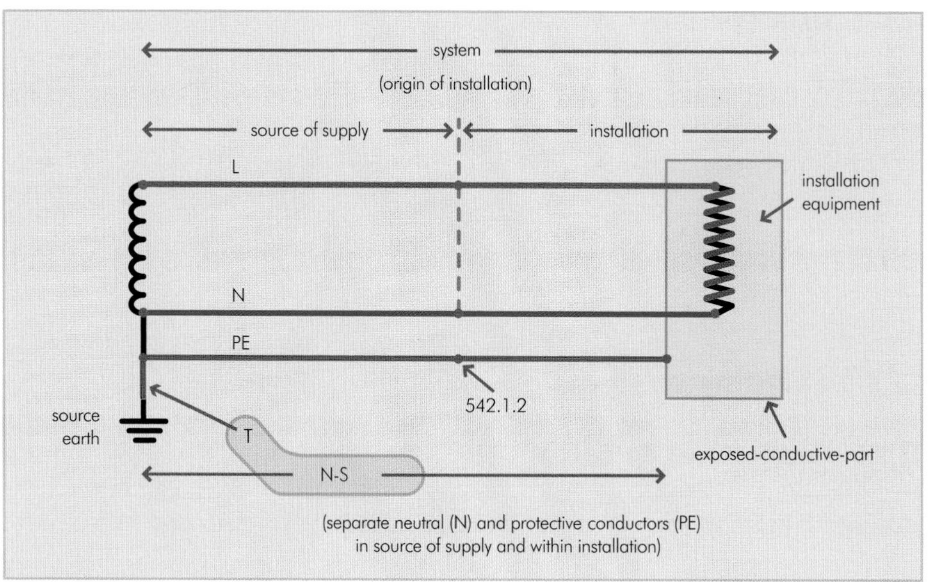

a TN-S system

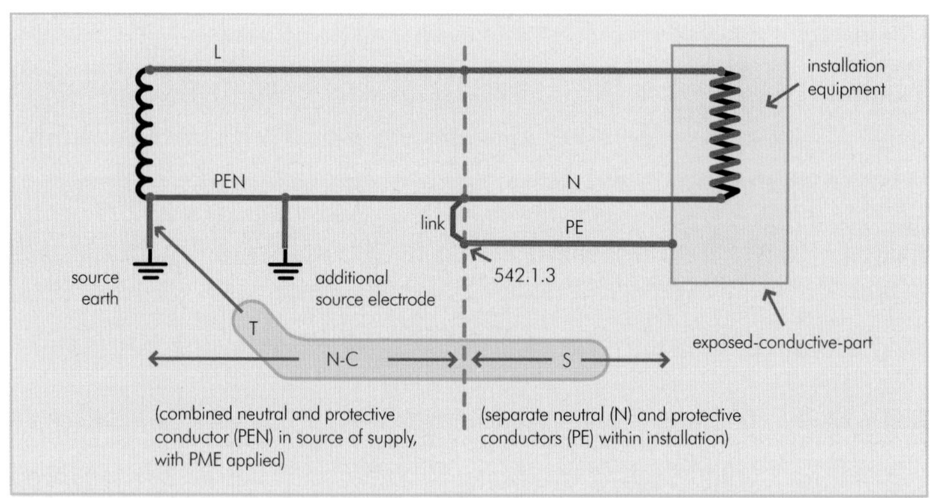

b TN-C-S system

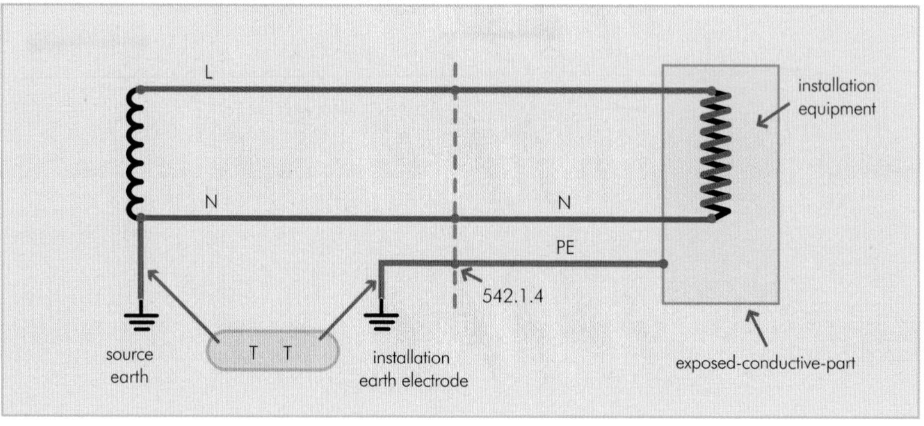

c TT system

11.1.2 TN-S system

A TN-S system (Figure 11.1a) has the neutral of the source of energy connected with Earth at one point only, at or as near as is reasonably practicable to the source, and the consumer's earthing terminal is typically connected to the metallic sheath or armour of the distributor's service cable into the premises or to a separate protective conductor of, for instance, an overhead supply. 542.1.2

11.1.3 TN-C-S system

A TN-C-S system (Figure 11.1b) has the supply neutral conductor of the distribution main connected with Earth at source and at intervals along its run. This is usually referred to as protective multiple earthing (PME). With this arrangement the distributor's neutral conductor is also used to return earth fault currents arising in the consumer's installation safely to the source. To achieve this, the distributor will provide a consumer's earthing terminal which is linked to the incoming neutral conductor. 542.1.3

PME sources offer a number of advantages over TN-S, among which are:

1 The connection with Earth is generally considered more reliable, as it does not depend upon factors such as the continuity of armouring
2 The PEN conductor, being of copper or aluminium, provides a lower impedance return path for earth fault currents than (say) steel armour or a lead sheath
3 Use of the neutral as a combined neutral and protective conductor represents a material saving
4 The PEN conductor has multiple connections to Earth along its run.

Main protective equipotential bonding is an important requirement for installations supplied from TN-C-S (PME) systems, in view of the risk of danger in the event of an open circuit neutral fault on the distributor's low voltage network. Regulation 25(2) of the ESQCR prevents a distributor from providing a PME supply to premises with inadequate bonding because the installation would not comply with BS 7671. 411.3.1.2

11.1.4 TT system

A TT system (Figure 11.1c) has the neutral of the source of energy connected as for TN-S, but no facility is provided by the distributor for the consumer's earthing.

With TT systems, the consumer must provide his or her own connection to Earth, i.e. by installing a suitable earth electrode local to the installation. The circumstances in which a distributor will not provide a means of earthing for the consumer are usually where the distributor cannot guarantee the earth connection back to the source, e.g. a low voltage overhead supply, where there is the likelihood of the earth wire either becoming somehow disconnected or even being stolen. A distributor also might not provide means of earthing for certain outdoor installations, e.g. a construction site temporary installation, leaving it to the consumer to make suitable and safe arrangements for which they are fully responsible. The electricity distributor is required to make available his supply neutral or protective conductor for connection to the consumer's earth terminal, unless this is inappropriate for reasons of safety (Regulation 24). Construction site, farm or swimming pool installations might be inappropriate unless additional precautions are taken, such as an additional earth electrode (see Chapter 14). 542.1.4

11.2 HV supplies

Large consumers may have one or more HV/LV transformers dedicated to their installation and installed adjacent to or within their premises. In such situations the usual form of system earthing is TN-S. It is not necessary for the star (neutral) point of the transformer LV secondary to be brought out at the transformer cable box in order to earth this point.

The usual and most convenient means of earthing the source neutral is for the connection to Earth to be made at the first accessible position in the LV system at which the neutral is terminated, i.e. LV feeder pillar or main LV switchboard.

Fig 2.2 Regulation 8(4) of the ESQCR prohibits TN-C systems, i.e. combined neutral and protective conductors, in a consumer's installation. However, if the person operating an extensive low voltage network on a site can be described as a distributor as defined by the ESQCR, and not a consumer, then he may operate a TN-C cable system to supply individual premises (installations). Each individual premises must be TN-C-S. If a TN-C (PME) distribution system is used, the requirements for protective multiple earthing laid down in Part II of the ESQCR for distributors must be closely followed, treating each building as if it housed a separate consumer.

542.1.8 In any premises where there are a number of buildings each having its own low voltage installation, care must be taken over the earthing of any circuit having its origin in one installation and which also runs into one or more other buildings. Unless suitable precautions are taken, the protective conductor (e.g. metallic cable sheath) of the circuit may be subjected to earth fault conditions exceeding its capability. Such a circuit may, for example, be concerned with a fire or security system.

542.2 11.3 Earth electrodes

542.2.1 11.3.1 Types of earth electrode

A wide variety of types of earth electrode are recognised by BS 7671, including:

1 earth rods or pipes
2 earth tapes or wires
3 earth plates
4 underground structural metalwork embedded in foundations
5 welded metal reinforcement of concrete (except pre-stressed concrete) embedded in the ground
6 lead sheaths and other metal coverings of cables, and
7 other suitable underground metalwork.

The most suitable type for a particular system will depend upon a number of factors, the single most important of these being the soil resistivity of the ground. If earth rods are to be driven it may be necessary to go down several metres before a good conductive layer is reached, especially where the water table is low. Rods can only be as effective as the contact they make with the surrounding material. Thus, they should be driven into virgin ground, not disturbed (backfilled) ground. Where it is necessary to drive two or more rods and connect them together to achieve a satisfactory result, the separation between rods should be at least equal to their driven depth to obtain maximum advantage from each rod.

In some locations low soil resistivity is found to be concentrated in the topsoil layer, beneath which there may be rock or other impervious strata which prevent the deep driving of rods, or a deep layer of high resistivity. Only a test or known information about the ground can reveal this kind of information. In such circumstances, the installation of copper earth tapes, or pipes or plates, would be most likely to provide a satisfactory earth electrode.

Closely spaced buildings may sometimes make it difficult to find ground suitable for driving an earth electrode. Electrodes which employ suitable structural or other underground metalwork, or the metal reinforcement of concrete embedded in the ground, may then be of particular advantage.

Whatever form an earth electrode takes, the possibility of soil drying and freezing, 542.2.2
and of corrosion, must be taken into account. Preferably, testing of an earth electrode 542.2.3
should be carried out under the least favourable conditions, i.e. after prolonged dry GN3
weather.

11.3.2 Pipes that must not be used as an earth electrode 542.2.4

The following must not be used as an earth electrode:

1. A metallic pipe for gases or flammable liquids, such as an oil feed pipe from an outdoor tank to a premises employing oil-fired heating or a gas supply pipe. The risk is that earth fault currents could cause a fire or explosion
2. The metallic pipe of a water utility supply. The risk here is that the supply company replace the pipe with a plastic pipe at some time in the future, leaving the electrical installation unearthed
3. Other metallic water pipework, unless its use as an earth electrode has been considered and precautions have been taken against its future removal. Such precautions might include labelling. Once again, the risk is one of removal leaving the electrical installation unearthed.

11.3.3 Use of a cable as an earth electrode 542.2.5

The unsleeved armour or lead sheath of a cable could be considered for use as an earth electrode because it may well be in good electrical contact with the general mass of Earth. However, the installation designer should consider the following:

1. Precautions should be taken to avoid deterioration by corrosion during the life of the electrical installation. Protective conductor currents can result both from fault currents and from equipment making use of the earth connection for filters, such as IT equipment. Such currents can cause corrosion
2. The consent of the owner of the cable must be obtained before his cable is used as an earth electrode
3. Arrangements must be in place so that the owner of the electrical installation is warned of any changes to the cable which might affect its suitability as an earth electrode. For example, the cable might be removed or replaced with a cable with an overall outer sheath resulting in the electrical installation being unearthed.

11.3.4 Additional earth electrode at the origin of an installation

BS 7671:2008 includes, in Regulation 411.4.2, a note that in TN systems the PE and 411.4.2 Note
PEN conductors may additionally be connected to Earth, such as at the point of entry into a building. See section 14.5 of this Guidance Note.

Further information on earthing principles and practice can be found in BS 7430 *Code of Practice for Earthing*.

11.4 Earthing conductor

11.4.1 Definition and purpose

Part 2 The *earthing conductor* of an electrical installation is the protective conductor that connects the main earthing terminal (MET) of the installation to the means of earthing, for example the earth connection provided by the distributor or an earth electrode, as illustrated in Figure 11.2. There is generally only one such conductor in each installation.

▼ **Figure 11.2**
Earthing conductor connecting an installation to its means of earthing

542.4.1

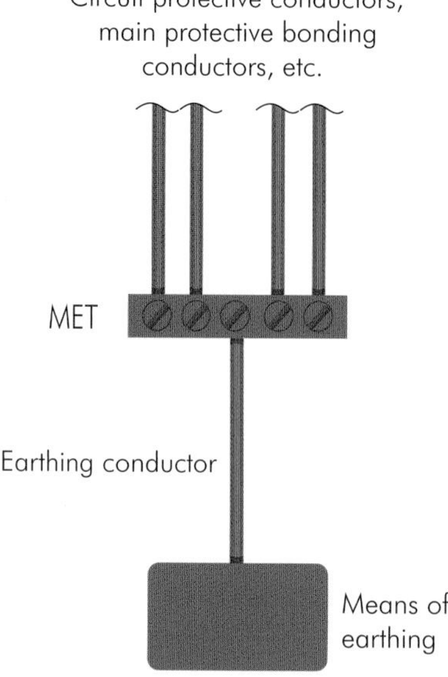

542.3.1 542.3.2 The earthing conductor of an installation forms part of the earth fault current loop. It must therefore be adequately sized for the maximum prospective fault current and, particularly where buried partly in the ground, be of suitable material and adequately protected against mechanical damage and corrosion.

542.4.2 A means must be provided to disconnect the earthing conductor in order to facilitate measurement of the external earth fault loop impedance. The means must be in an accessible position and may be in the form of a disconnectable link combined with the MET, or a joint capable of disconnection only by means of a tool (e.g. spanner or screwdriver).

11.4.2 Earthing conductor cross-sectional area

542.3.1 Table 54.7 Table 54.1 The cross-sectional area (csa) of an earthing conductor is determined in basically the same way as for other protective conductors, i.e. either by selection from Table 54.7 or by calculation – see Chapter 12. However, in addition, for a buried earthing conductor the requirements of Table 54.1 must be met, to ensure it is mechanically strong and of sufficient csa to be resistant to corrosion, and, for a TN-C-S (PME) supply, the earthing conductor must be no smaller than the main protective bonding conductors Table 54.8 as required by Table 54.8.

11.5 Main earthing terminal or bar

Every installation must have a main earthing terminal or bar for connection of the following to the earthing conductor: Part 2 542.4.1

1 the circuit protective conductors
2 the protective bonding conductors
3 unctional earthing conductors (if required)
4 lightning protection system bonding conductor (if any).

This is illustrated in Figure 11.2.

Note that item 4 requires a lightning protection system to be bonded to the main earthing terminal and the system's down conductors to be still connected to their own earth electrodes, which must be tested before the main protective bonding connection is made (see BS EN 62305 *Protection against lightning* for further guidance). 411.1.1.2

For a larger installation having an LV switchboard, the main earth bar will often be located within the switchboard. An alternative arrangement, sometimes used, is for the earth bar to be fixed external to the switchboard, i.e. mounted on insulating supports on the switchroom wall. This makes for ready inspection of both its condition and the connections of the protective conductors. Disconnection of protective conductors for testing, where necessary, is also made easier by this arrangement. 542.4.2 543.3.3

A main earth bar needs to be adequately sized for mechanical and electrical purposes. If the bar is drilled to receive bolts for the connection of protective conductors, its minimum cross-sectional area (csa), i.e. taking into account the diameter of the largest hole drilled in it, should be not less than the csa of the earthing conductor (assuming they are of the same material).

A warning label (to BS 951) with the words 'Safety Electrical Connection – Do Not Remove' (Figure 11.3) should be permanently fixed at or near the main earthing terminal where this is separate from the main switchgear. 514.13.1

▼ **Figure 11.3**
BS 951 earthing and bonding warning label

11.6 Functional earthing

Confusion sometimes arises over the difference between protective earthing (PE) and functional earthing (FE). Protective earthing, as the name suggests, is provided for the safety of persons, livestock and property, and for the most part is what this Guidance Note is concerned with. A functional earth is, however, only provided to enable equipment to operate correctly. At no time does the FE offer protection to either the user or the equipment. Part 2

543.5.1 Where earthing for combined protective and functional purposes is required, the
542.1.5 requirements for protective measures must take precedence. The earthing arrangements may be used jointly or separately for protective and functional purposes, according to the requirements of the installation.

Examples of functional earthing are:

1 to provide a 0 (zero) volt reference point
2 to enable an electromagnetic screen to be effective
3 to provide a signalling path for some types of communications equipment.

542.4.1 The most common use of a functional earth is for telecommunication purposes. It is permissible for the FE to be terminated onto the electrical installation main earthing terminal or bar. The wiring used will normally be of copper and at least 1.5 mm^2 in size.
Table 51 It should be coloured cream, as stated in Table 51 of BS 7671, and at the main earthing terminal it should have a label (or embossed sheath) marked 'Telecommunication Functional Earth'. Electrical installers who come across such functional earths, which are particularly common in commercial buildings, should not interfere with these connections.

Functional earthing may also be required for other equipment and should be coloured cream. The connection to earth should again be made to the main earthing terminal or bar and be clearly labelled as to its purpose.

Circuit protective conductors

12

12.1 Introduction

A circuit protective conductor (cpc) forms part of the earth fault loop, so that, in the event of an earth fault in the circuit with which it is associated, sufficient current flows to operate the protective device in the required time. The main function of a cpc is therefore to carry the earth fault current without damage either to itself or to its surroundings, e.g. insulation. Part 2

Figure 12.1 illustrates the path the earth fault current will take in the case of a TN-C-S system.

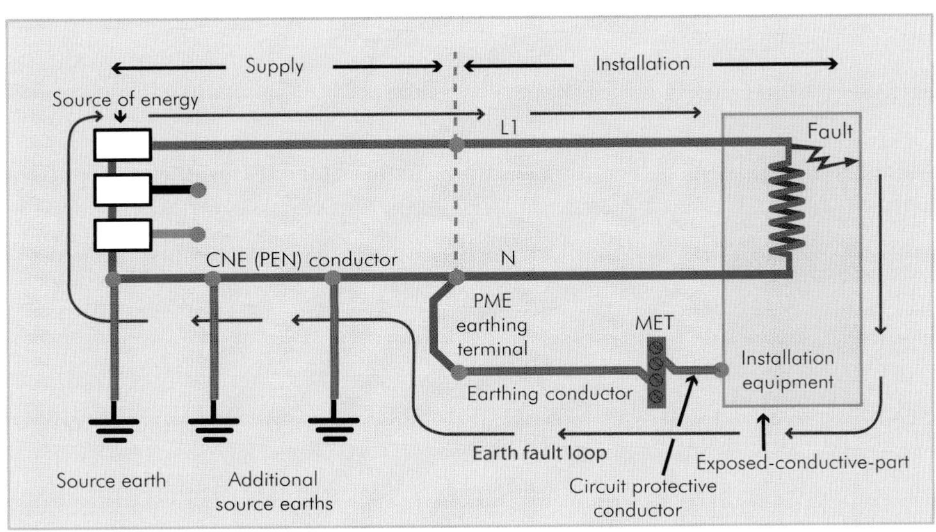

▼ **Figure 12.1**
Earth fault current path in a TN-C-S system

12.2 Sizing of circuit protective conductors

BS 7671 provides two methods for sizing protective conductors: 543.1.1 543.1.3 543.1.4 Table 54.7

i Selection from Table 54.7 of BS 7671, reproduced here as Table 12.1
ii Calculation using the adiabatic equation.

Some earthing conductors can be sized similarly; see 11.4.2.

12.2.1 Selection (from Table 12.1)

The easier method is to determine the protective conductor size from Table 54.7 of BS 7671 reproduced here as Table 12.1, but this may produce a larger size than is necessary, since it employs a simple relationship to the cross-sectional area of the line conductor(s). Oversizing is particularly liable to happen where the size of the live conductors of the circuit has been increased in order to satisfy grouping or voltage drop.

Table 54.7 ▼ **Table 12.1** Minimum cross-sectional area of protective conductor in relation to the cross-sectional area of associated line conductor

Cross-sectional area of line conductor, S	Minimum cross-sectional area of the corresponding protective conductor	
	If the protective conductor is of the same material as the line conductor	If the protective conductor is not of the same material as the line conductor
(mm²)	(mm²)	(mm²)
$S \leq 16$	S	$\frac{k_1}{k_2} \times S$
$16 < S \leq 35$	16	$\frac{k_1}{k_2} \times 16$
$S > 35$	$\frac{S}{2}$	$\frac{k_1}{k_2} \times \frac{S}{2}$

Notes:

Table 43.1 k_1 is the value of k for the line conductor, selected from Table 43.1 of BS 7671 according to the materials of both conductor and insulation

Tables 54.2 to 54.6 k_2 is the value of k for the protective conductor, selected from Tables 54.2 to 54.6 of BS 7671, as applicable.

It should also be noted that, except for 1 mm², twin and earth cables do not comply with Table 54.7 of BS 7671, as the protective conductor is smaller than the live conductors. For circuits that employ these cables, the adequacy of the protective conductor must be checked by using the adiabatic equation method of sizing described below.

Application of the adiabatic equation method will in many instances result in a protective conductor of smaller csa than that of the live conductors of the associated circuit, which is quite acceptable.

12.2.2 Adiabatic calculation

543.1.3 The adiabatic equation of Regulation 543.1.3 is so-called because it assumes all heat generated is retained in the cable for the period of the fault and no account is taken of heat loss by conduction, convection or radiation. Thus, it errs on the safe side.

543.1.3 The cross-sectional area, where calculated, shall be not less than the value determined by the following formula or shall be obtained by reference to BS 7454:

$$S = \frac{\sqrt{I^2 t}}{k}$$

Note: This equation is an adiabatic equation and is applicable for disconnection times not exceeding 5 s.

where:

- S is the nominal cross-sectional area of the conductor in mm^2
- I is the value in amperes (rms for a.c.) of fault current for a fault of negligible impedance, which can flow through the associated protective device, due account being taken of the current limiting effect of the circuit impedances and the limiting capability (I^2t) of that protective device
- t is the operating time of the disconnecting device in seconds corresponding to the fault current I amperes
- k is a factor taking account of the resistivity, temperature coefficient and heat capacity of the conductor material, and the appropriate initial and final temperatures.

Values of k for protective conductors in various use or service are as given in Tables 54.2 to 54.6. The values are based on the assumed initial and final temperatures indicated in each table.

Where the application of the formula produces a non-standard size, a conductor having the nearest larger standard cross-sectional area shall be used.

In order to apply the adiabatic equation, either:

i values of I and t, or
ii (I^2t)

must be determined.

Where the manufacturer of a device provides the (I^2t) for a given fault level, S can be calculated after looking up k in Tables 54.2 to 54.6 as appropriate. Otherwise the prospective earth fault current I can be calculated using Ohm's law:

$$I = U_0/Z_s \text{ (A)}$$

where:

- U_0 is the nominal voltage to Earth in volts
- Z_s is the earth fault loop impedance in ohms at the farthest point of the circuit being protected.

The calculation of Z_s is described in Appendix B, for an assumed (design) protective conductor size.

When I has been determined, t is obtained for standard devices from the time/current characteristics in Appendix 3 of BS 7671 or from manufacturers' data. Appx 3

The adiabatic equation of Regulation 543.1.3 given above can then be used to confirm the minimum acceptable size of protective conductor. Some iteration may be necessary if the assumed size of the protective conductor is too small.

Appendix A provides maximum values of earth fault loop impedance for all the common overcurrent devices. If the value of earth fault loop impedance given in the tables for a particular device and particular size of circuit protective conductor is not exceeded, the requirement of the adiabatic equation will be met, as will the requirement for fault protection by automatic disconnection of supply.

IEE publication *Electrical Installation Design Guide: Calculations for Electricians and Designers* details these calculations and provides examples.

12.3 Provision of circuit protective conductors

543.2 There are a number of types of circuit protective conductor (cpc) in common use, including copper conductors, metal wiring enclosures such as a rigid conduit, trunking, ducting or riser system, as well as the metal covering of cables, such as the sheath or armouring.

There are only three instances in BS 7671 where a separate circuit protective conductor is prescribed:

543.2.1 **1** through flexible or pliable metal conduit
543.7 **2** subject to specified criteria, in a final circuit intended to supply equipment producing protective conductor current in excess of 10 mA in normal service
543.2.7 **3** to connect the earthing terminal of an accessory to metal conduit, trunking or ducting.

Apart from these instances, there are circumstances for which it is recommended that consideration be given to the provision of a separate circuit protective conductor if metal conduit, trunking or ducting is used. These include:

1. industrial kitchens, laundries and other 'wet areas'
2. locations where chemical attack or corrosion of the metal wiring enclosure or cable sheath may be expected
3. circuits exceeding 160 A rating
4. any location where the integrity of the metal-to-metal joints in the conduit/trunking installation cannot be ensured over the life of the installation.

Note: Where full information concerning correct and reliable earthing is provided by the manufacturer, such as for busbar trunking and powertrack installations, the system may be used up to its maximum rating.

411.3.1.1 It is stressed that where a separate circuit protective conductor is installed, any metal
543.3 cable management system must still be properly constructed, provide good continuity throughout its run and be earthed. The metalwork is still an exposed-conductive-part containing live conductors and might otherwise give rise to danger in the event of an insulation fault.

543.1.2 Where a cpc is used to serve a number of circuits, it must meet the requirement for the most onerous duty. If the cross-sectional area (csa) of the cpc is determined by selection, then the csa relating to the larger or largest line conductor of all the circuits which are served by the common cpc must be used. However, where the calculation option is used, the most onerous values of fault current and disconnection time should be used.

It should be noted that the lower values of fault current may not necessarily produce the lowest I^2t values; normally the reverse is the case because a low fault current leads to a relatively high disconnection time.

12.4 High protective conductor currents 543.7

12.4.1 Introduction

High protective conductor currents, that is protective conductor currents exceeding say 10 mA, can present a risk of electric shock should the protective conductor become disconnected.

A protective conductor current of 10 mA equates to a loop impedance of 23 000 ohms (230 V/10 mA). If the connection of the protective conductor with earth is lost, a person touching exposed-conductive-parts of the equipment and in contact with earth will complete an earth fault loop with an impedance of about 24 000 ohms – see Figure 12.2. This will result in a potential body or touch current of 9.6 mA.

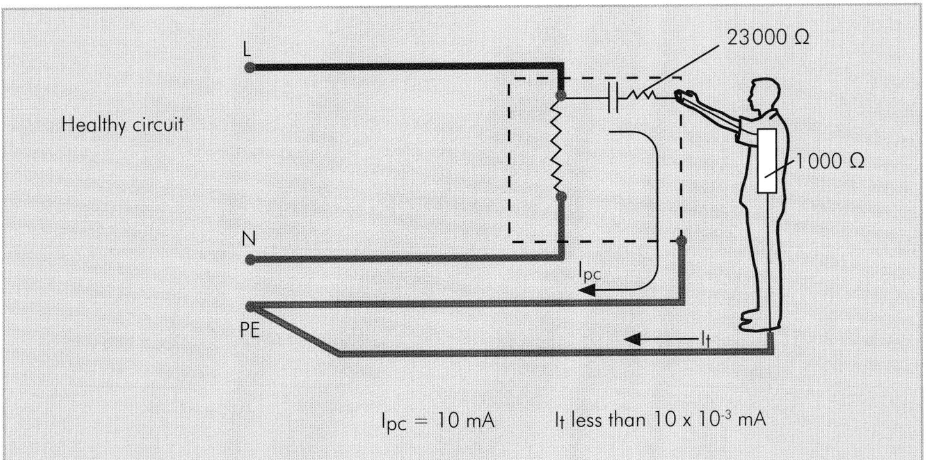

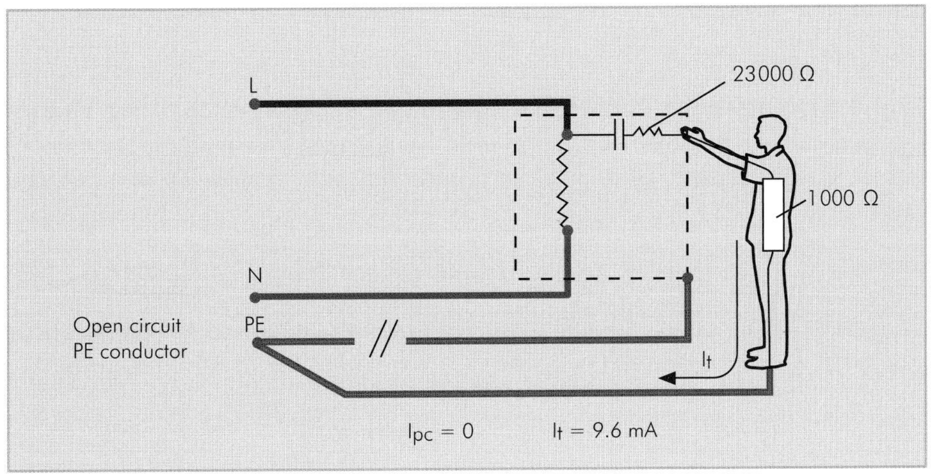

▼ **Figure 12.2** Touch currents with open circuit protective conductor

This current is likely to be on the border of zones AC-2 and AC-3 of Figure 2.1 and Table 2.1 of Chapter 2. AC-2 is the zone where perception and involuntary muscular contractions are likely but usually no harmful physiological effects; zone AC-3 is where there may be strong involuntary muscular contractions, difficulty in breathing, etc.

12.4.2 Equipment with protective conductor current exceeding 3.5 mA

543.7.1.1 Regulation 543.7.1.1 requires equipment having high protective conductor current exceeding 3.5 mA but not exceeding 10 mA to be either permanently connected to the fixed wiring without the use of a plug and socket-outlet, or connected by means of an industrial plug and socket-outlet to BS EN 60309-2.

These limits are related to those of the standard for household and similar appliances and that for IT equipment. The standard for household and similar appliances, BS EN 60335, generally requires equipment protective conductor currents to be less than 3.5 mA.

The standard for IT equipment, BS EN 60950-1, requires 'pluggable type A' equipment, that is equipment to be connected with a standard 13 A plug, to have a protective conductor current not exceeding 3.5 mA. Where the protective conductor current exceeds 3.5 mA the standard requires the equipment to be permanently connected or connected via an industrial plug and socket. This is called pluggable type B equipment. There is also a requirement in BS EN 60950 for a label as shown below for equipment with touch current exceeding 3.5 mA.

WARNING
HIGH LEAKAGE CURRENT.
EARTH CONNECTION ESSENTIAL
BEFORE CONNECTING SUPPLY.

12.4.3 Equipment with protective conductor current exceeding 10 mA

543.7.1.2 Regulation 543.7.1.2 requires equipment having an earth leakage current exceeding 10 mA to be either permanently connected to the fixed wiring (with minimum protective conductor requirements), connected by an industrial plug and socket-outlet (with minimum flex protective conductor csa), or provided with a protective conductor that is monitored.

12.4.4 Circuits with protective conductor current exceeding 10 mA

543.7.1.3 Regulation 543.7.1.3 requires circuits to equipment having leakage currents exceeding 10 mA to have a high integrity protective conductor or equivalent arrangements made.

Circuits supplying IT equipment may have protective conductor currents exceeding 10 mA as may many other circuits, including luminaire circuits.

As discussed above, BS EN 60950-1 requires 'pluggable type A' equipment, that is equipment to be connected with a standard 13 A plug, to have a protective conductor current not exceeding 3.5 mA. In practice, flat screen PCs may have a protective conductor current of the order of 1 mA.

Luminaire high-frequency ballasts have RFI filter networks which will include capacitor connections to earth. Typically, the protective conductor current per ballasted luminaire is up to 1 mA. (BS EN 60598-1:1989 allows 1 mA for up to 1 kVA input; BS EN 60598-1:2008 allows up to 3.5 mA for luminaires with a supply current of 7 A or less (clause 10.3).)

12.4.5 Socket-outlet final circuits 543.7.2.1

Figures 12.3 and 12.4 illustrate typical ring and radial final circuit arrangements.

▼ **Figure 12.3** Ring final circuit supplying socket-outlets (total protective conductor current exceeding 10 mA)

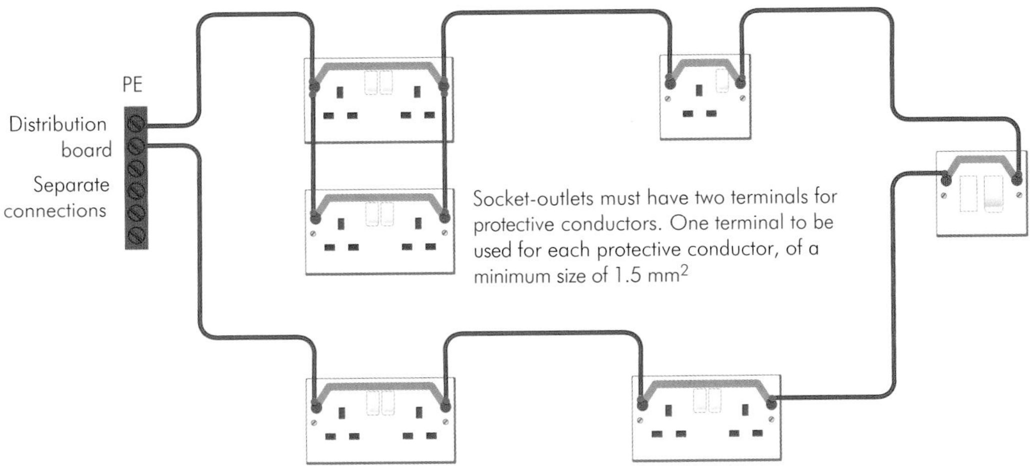

▼ **Figure 12.4** Radial final circuit supplying twin socket-outlets (total protective conductor current exceeding 10 mA), with duplicate protective conductor

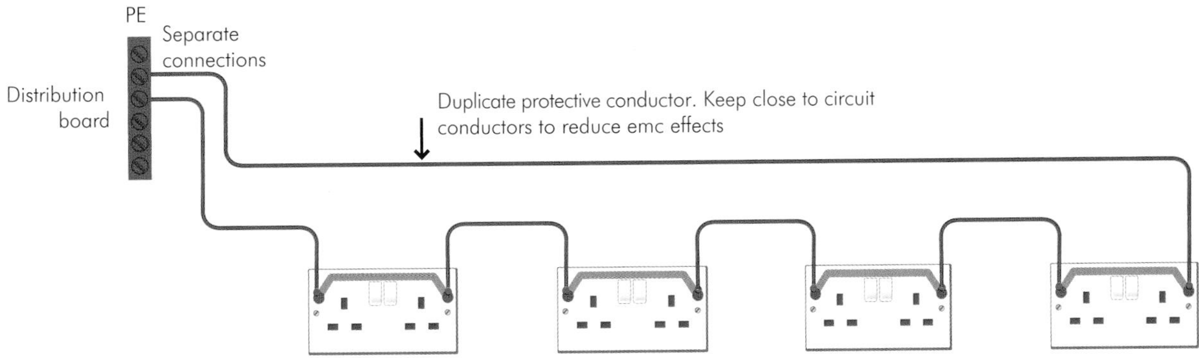

Protective equipotential bonding 13

13.1 Introduction

The purpose of earthing the exposed-conductive-parts of an installation is to ensure that, in the event of a fault (line conductor to an exposed-conductive-part), sufficient fault current flows to operate the disconnection device (fuse, circuit-breaker, RCD). Earthing exposed-conductive-parts also reduces the touch voltage (U_t) between these and extraneous-conductive-parts during a fault. 411.3.1.1

The purpose of protective equipotential bonding is to further reduce the touch voltage between exposed-conductive-parts and extraneous-conductive-parts in the event of: Part 2 411.3.1.2

i a fault on the installation
ii an open circuit PEN conductor in a PME supply.

(The multiple earthing required by the Electricity Safety, Quality and Continuity Regulations 2002 for TN-C-S systems also significantly reduces touch voltages. This is discussed further in Chapter 14.)

13.2 Main protective equipotential bonding

13.2.1 Touch voltages

The effect of applying main protective equipotential bonding is most noticeable in TT systems. Consider Figure 13.1.

The touch voltage in the event of a fault with main protective bonding installed is given by:

$$U_t = I_f\,(R_2)$$

The touch voltage (within the installation) in the event of a fault with no main protective bonding is given by:

$$U_t = I_f\,(R_2 + R_A)$$

I_f is given by:

$$I_f = \frac{U_0}{R_2 + R_1 + R_A + R_B}$$

▼ **Figure 13.1**
TT system

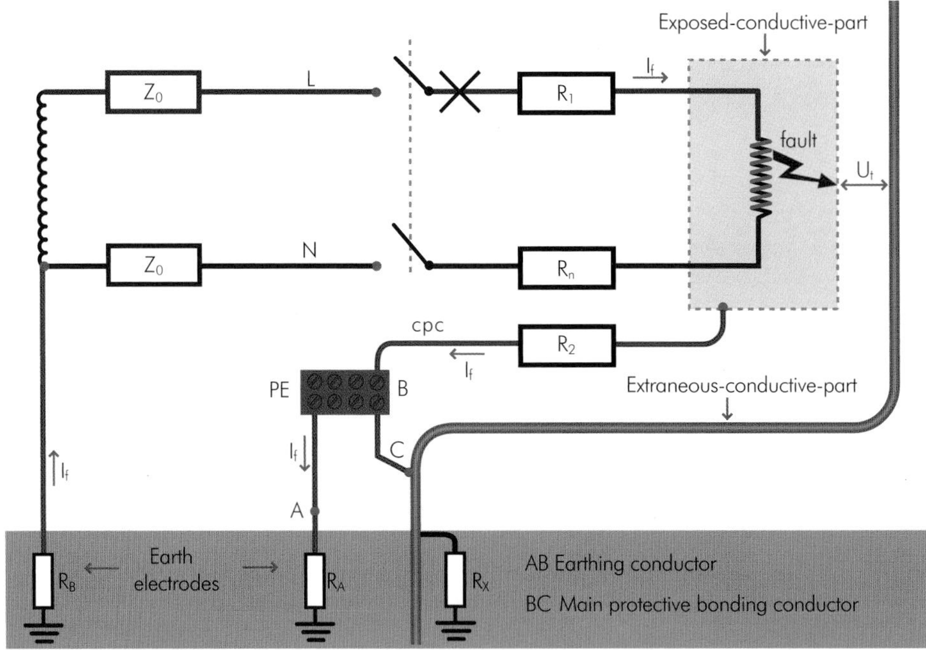

Assume $U_0 = 230$ V, $R_B = 20\ \Omega$, $R_A = 100\ \Omega$, $R_1 = 0.5\ \Omega$, $R_2 = 0.5\ \Omega$

Then

$$I_f = \frac{230}{0.5 + 0.5 + 100 + 20}$$

$$= 230/121 = 1.9\text{ A},$$

giving U_t with no bonding = 191 V

and U_t with bonding = 0.95 V

For a TN-C-S installation consider Figure 13.2.

The touch voltage in the event of a fault with main protective bonding installed is given by:

$$U_t = I_f\ (R_2)$$

The touch voltage (within the installation) in the event of a fault with no main protective bonding is given by:

$$U_t = I_f \left(R_2 + \frac{Z_e}{2} \right)$$

I_f is given by:

$$I_f = \frac{U_0}{R_2 + R_1 + Z_e}$$

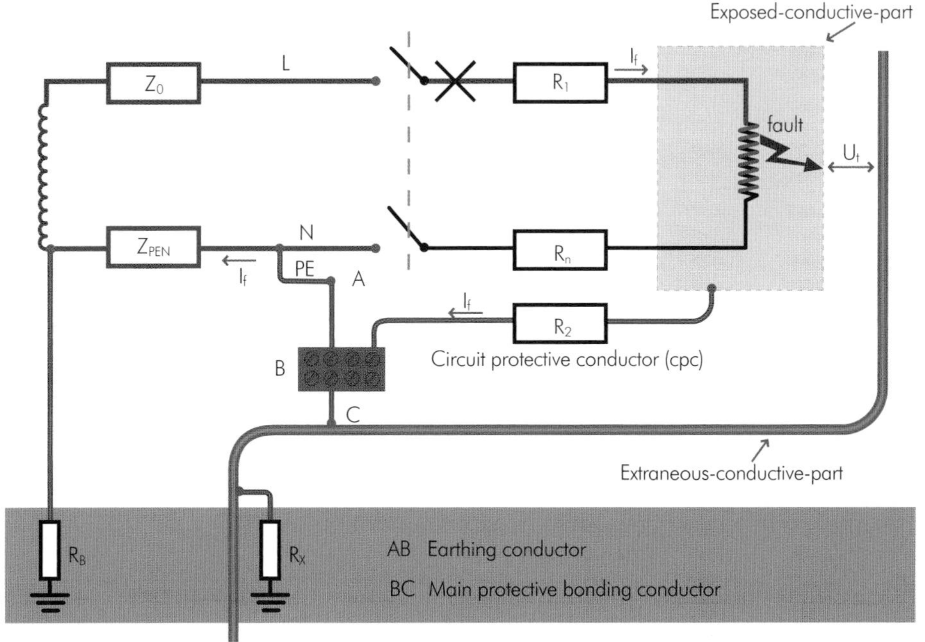

▼ **Figure 13.2**
TN-C-S system

Assume $U_0 = 230$ V, $Z_e = 0.35\ \Omega$, $R_1 = 0.5\ \Omega$, $R_2 = 0.5\ \Omega$

Then

$$I_f = \frac{230}{0.5 + 0.5 + 0.35}$$

$$= 230/1.35 = 170 \text{ A},$$

giving U_t with no bonding = 115 V

and U_t with bonding = 85 V

13.2.2 Sizing of main protective bonding conductors 544.1

TN-S and TT systems

For TN-S and TT systems, the current flowing in a protective bonding conductor will be minimal. The main bonding conductors must have a cross-sectional area (csa) of not less than half the cross-sectional area required for the earthing conductor, with a minimum conductor size of 6 mm^2 (for mechanical strength). The csa need not exceed 25 mm^2.

TN-C-S systems

For PME conditions (TN-C-S systems) the minimum cross-sectional area of main protective bonding conductors is given in Table 54.8 of BS 7671. These minimum sizes are related to the cross-sectional area of the supply neutral conductor. In the event of an open circuit PEN conductor, load currents will return to the earthed neutral point of the supply via main bonding conductors and extraneous-conductive-parts. The current will be related to the load, which is loosely linked to the supply cable csa, so it is reasonable to apply the relationship in Table 54.8. Table 54.8

Whatever the type of system, it is recommended that the electricity distributor or supplier should be asked to confirm their agreement in writing to the proposed size(s) of main bonding conductors it is intended to install.

13.2.3 Installation of bonding conductors

Part 2 Main protective equipotential bonding is achieved by connecting all extraneous-
conductive-parts within a building to the main earthing terminal or bar of the
413.3.1.2 installation by main protective bonding conductors. There must not be any substantial*
extraneous-conductive-part within the building which is not bonded, or the principle of equipotential bonding is breached and the presence of such a part could be a danger, especially if it is simultaneously accessible with any exposed-conductive-part of the installation.

Main protective bonding conductors should be kept as short as practicable and be
544.1.2 routed to minimise the likelihood of damage or disturbance to them. The connections
to gas, water and other services entering the premises must be made as near as practicable to the point of entry of each service, on the consumer's side of any insulating section or insert at that point or any meter. Any substantial extraneous-conductive-part which enters the premises at a point remote from the main earthing terminal or bar must also be bonded to this terminal or bar.

Extraneous-conductive-parts should preferably be bonded using individual main protective bonding conductors. Alternatively, two or more such parts may share a bonding conductor, but where this arrangement is employed the conductor should be continuous, i.e. disconnection of the conductor from one extraneous-conductive-part must not interfere with or endanger the security of the bonding of the other part(s).

411.3.1.2 Where the installation is within a building which has exposed metallic structural parts,
544.1.2 then these parts must also be bonded if they are extraneous-conductive-parts, i.e.
the structural metalwork is in contact with the ground or other unbonded earthed metalwork. Where the structure concerned is assembled from components which are welded, bolted or even riveted together then the number of main bonding connections can be minimised; see Figure 13.3.

543.2.6 If doubt exists over the electrical continuity between the main bond and the parts of the structure in contact with Earth, it is recommended that a continuity test should be carried out to determine the extent, if any, of cross-bonding between main structural components that may be required in order to achieve this continuity.

When carrying out such tests the possibility of unsuitable or fortuitous earth paths must be considered.

514.13.1 A warning label with the words:

Safety Electrical Connection – Do Not Remove

should be permanently fixed in a visible position, at or near the point of connection of every main protective bonding conductor to an extraneous-conductive-part (see Figure 11.3 in section 11.5 of this Guidance Note).

---

* Excludes any part of small dimensions that cannot be readily or reliably bonded and which cannot be gripped or contacted by a major surface of the human body, e.g. does not exceed 50 mm × 50 mm.

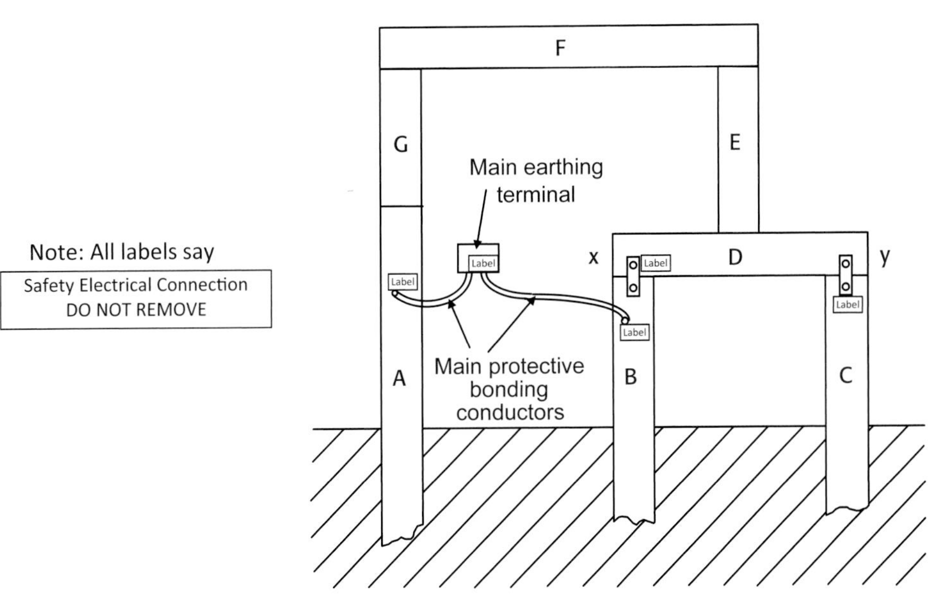

▼ **Figure 13.3** Main bonding of structural steelwork

Notes:

- A, B and C are extraneous-conductive-parts. A and B are bonded directly to the main earthing terminal, C is bonded by reliable connections X and Y.
- E, F and G, although not reliably connected to the main earthing terminal, do not require main bonding. They may require supplementary bonding if accessible in areas of increased shock risk.

13.3 Supplementary equipotential bonding

13.3.1 Introduction

Supplementary equipotential bonding is required by BS 7671 to be provided in the following situations:

i where the conditions for automatic disconnection cannot be fulfilled 411.3.2.6

ii for additional protection in some installations and locations of increased shock risk. Part 7

Where applicable, Regulation 415.2.1 requires supplementary equipotential bonding to connect together all the *simultaneously accessible* exposed-conductive-parts of fixed equipment and extraneous-conductive-parts including, where practicable, the main metallic reinforcement of constructional reinforced concrete. The equipotential bonding system must be connected to the protective conductors of all equipment including those of socket-outlets. 415.2.1

The particular requirements of Part 7 may include requirements additional to those in Regulation 415.2.1. For example, for a location containing a swimming pool, all the extraneous-conductive-parts (not just the simultaneously accessible) shall be connected to the protective conductors of exposed-conductive-parts of equipment in the zones. 702.415.2

13.3.2 Sizing of supplementary bonding conductors

Supplementary bonding conductors are unlikely to carry any significant currents as they are supplementary to the main bonding. Mechanical considerations are usually the limiting factor. See Table 13.1.

544.2 ▼ **Table 13.1** Supplementary bonding conductors

Size of circuit protective conductor (mm^2)	Minimum cross-sectional area of supplementary bonding conductors (mm^2)					
	Exposed-conductive-part to extraneous-conductive-part		Exposed-conductive-part to exposed-conductive-part		Extraneous-conductive-part to extraneous-conductive-part*	
	mechanically protected	not mechanically protected	mechanically protected	not mechanically protected	mechanically protected	not mechanically protected
	1	2	3	4	5	6
1.0	1.0	4.0	1.0	4.0	2.5	4.0
1.5	1.0	4.0	1.5	4.0	2.5	4.0
2.5	1.5	4.0	2.5	4.0	2.5	4.0
4.0	2.5	4.0	4.0	4.0	2.5	4.0
6.0	4.0	4.0	6.0	6.0	2.5	4.0
10.0	6.0	6.0	10.0	10.0	2.5	4.0
16.0	10.0	10.0	16.0	16.0	2.5	4.0

* If one of the extraneous-conductive-parts is connected to an exposed-conductive-part, the bond must be no smaller than that required for bonds between exposed-conductive-parts and extraneous-conductive-parts – column 1 or 2.

13.3.3 Supplementary bonding of fixed appliances

544.2.5 The circuit protective conductor (cpc) within the flexible cord supplying a fixed appliance may be used to provide the supplementary bonding connection to the appliance; see Figure 13.4.

▼ **Figure 13.4** Use of cpc in a short length of flexible cord for supplementary bonding

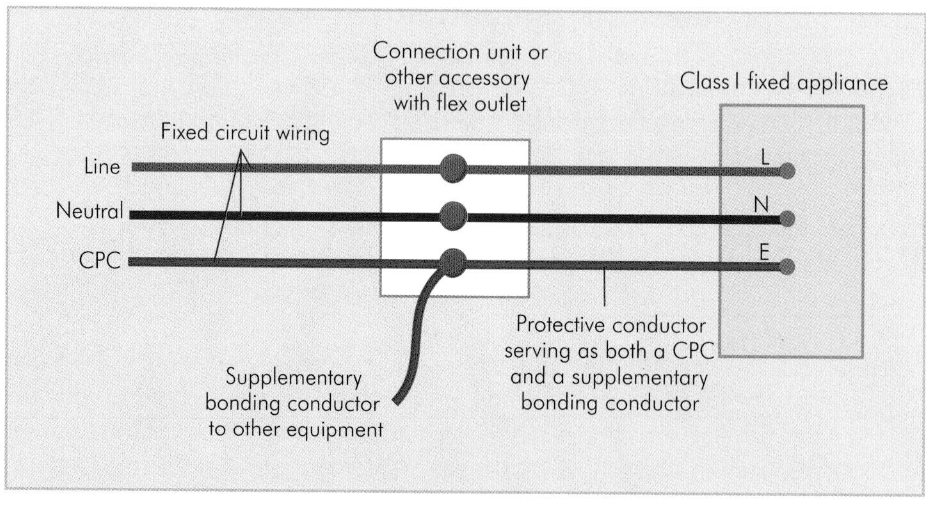

13.3.4 Resistance between supplementary bonded items

415.2.2 Regulation 415.2.2 imposes a limit on the ohmic value of the resistance (R) between supplementary bonded exposed-conductive-parts and extraneous-conductive-parts in a.c. systems, as given in the following equation:

$$R \leq 50\ V/I_a$$

where I_a is the operating current in amperes of the protective device:

- for RCDs, $I_{\Delta n}$.
- for overcurrent devices, the current causing automatic operation in 5 s.

Meeting these requirements for resistance (R) is unlikely to be a problem in practice; see section 8.3.3.

13.3.5 Warning labels

A warning label with the words: 514.13.1

Safety Electrical Connection – Do Not Remove

should be permanently fixed in a visible position, at or near the point of connection of every supplementary bonding conductor to an extraneous-conductive-part (see Figures 11.3 and 13.5), but see also the final paragraph of 13.5.

13.3.6 Locations where supplementary bonding is not required

If the installation meets the requirements for earthing and protective bonding, there is no specific requirement in BS 7671 to supplementary bond the following:

- locations containing a bath or shower, providing RCDs protect all the circuits of the location 701.415.2
- kitchen pipes, sinks or draining boards
- metal boiler pipework
- metal furniture in kitchens
- metal pipes to wash handbasins and WCs in domestic locations.

Note: Metal waste pipes in contact with Earth must be bonded to the main earthing terminal.

13.4 Example of supplementary bonding

Figure 13.5 shows two items of current-using equipment 'A' and 'B' and the circuit protective conductors of the circuits feeding them, together with two separate extraneous-conductive-parts 'C' and 'D'. In accordance with BS 7671 the circuit protective conductor and the extraneous-conductive-parts are connected to the main earthing terminal (E) of the installation.

Supplementary bonding is carried out by installing bonding conductors between the exposed-conductive-parts and the extraneous-conductive-parts, by making the supplementary bonding connections shown. The conductor connecting equipment 'A' and 'B' must comply with Regulation 544.2.1, while that between equipment 'B' and 544.2.1
extraneous-conductive-part 'C' must comply with Regulation 544.2.2. 544.2.2

The extraneous-conductive-part 'D' is connected either to current-using equipment A or B or to extraneous-conductive-part C, as indicated by the broken lines. Where it is connected to either item of current-using equipment, the bonding conductor is required to comply with Regulation 544.2.2. Alternatively, where the connection 544.2.2
is made instead to the extraneous-conductive-part C, compliance with Regulation 544.2.3
544.2.3 is required. It should be noted that, as permitted by Regulation 544.2.4, the 544.2.4
portion of extraneous-conductive-part between points F and G can be considered to be part of the supplementary bonding if it meets the requirements of Regulation 543.2.6
543.2.6.

It is not a requirement of BS 7671 to connect the supplementary bonding back to the main earthing terminal of the installation, although the locally bonded parts will be connected to this terminal by virtue of one or more circuit protective conductors and/or extraneous-conductive-parts as shown in Figure 13.5.

▼ **Figure 13.5**
Illustration of the application of supplementary bonding

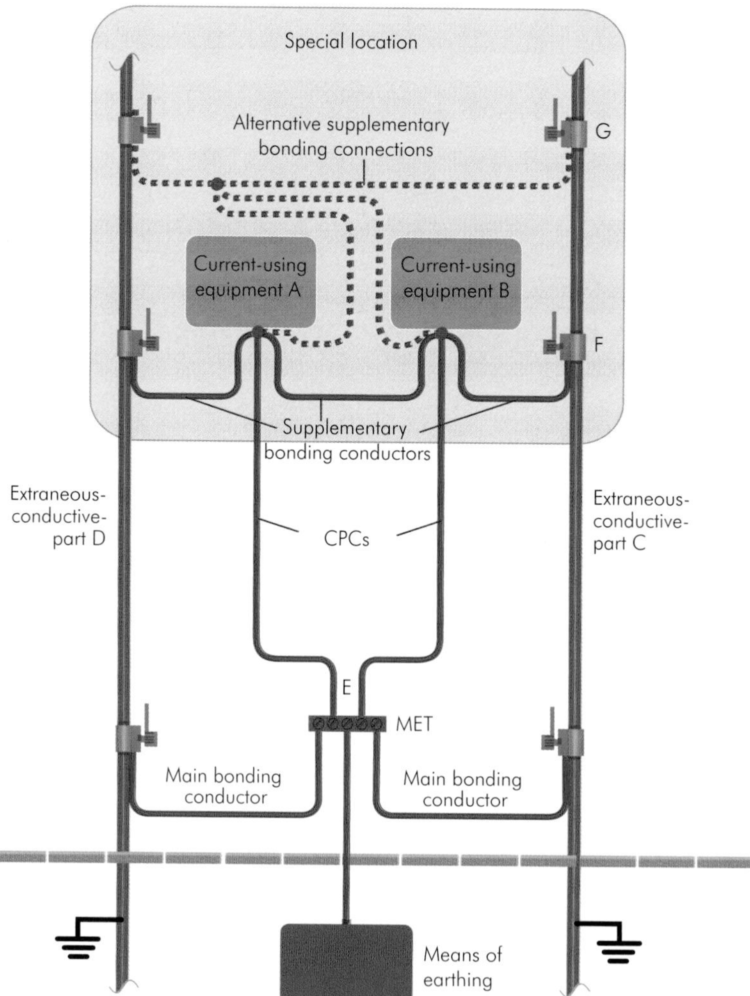

13.5 Extraneous-conductive-parts

In carrying out an assessment of what requires to be bonded the designer (or possibly the installer) may encounter difficulty in determining whether a part is actually an extraneous-conductive-part as defined, i.e. is liable to introduce a potential, generally Earth potential. In cases of doubt, a measurement of the resistance (R_x) between the conductive part concerned and the main earthing terminal of the installation should be made.

Part 2

The part can be considered not to be an extraneous-conductive-part if the resistance R_x is such that

$$\frac{U_0}{R_b + R_x} < I_b$$

where:

U_0 is the nominal voltage to Earth
R_b is the resistance of the human body
I_b is the value of the current through the human body which should not be exceeded.

The values of both R_b and I_b should be chosen from the data in DD IEC/TS 60479 (see Chapter 2) and are dependent on the conditions which are expected to arise in the installation concerned. For example, where the shock risk is likely to be from hand-to-hand contact, it is suggested that a suitable value to take for I_b would be 10 mA and for R_b 1 000 ohms, leading to a value of R_x of 22 000 ohms (for $U_0 = 230$ V), below which the part concerned is taken to be an extraneous-conductive-part.

If an acceptably high resistance is measured under dry conditions in a situation where at times moisture is to be expected, then either the test should be repeated under 'worst conditions', or supplementary bonding should be applied.

It is necessary to look a little further when deciding about the bonding of metalwork accessible to people or livestock outside the building and in contact with the ground. For example, an unbonded metal window frame inserted in a brick building is not likely to present a hazard, but when it is bonded it may well achieve a considerable rise in potential during the permitted 5 s clearance of a fault within the installation. This could be dangerous to a window cleaner outside the building, with wet hands and standing on damp ground or on a metal ladder. On the other hand, a metal window frame in a metal-clad building could introduce Earth potential into the location.

There is no single answer to the bonding of extraneous-conductive-parts, each situation has to be considered on its merits, and a decision made which, on balance, will provide the greater degree of safety.

The possibility of using the extraneous-conductive-part itself as the bonding conductor 543.2.6
permits a considerable economy in installation. There is, for example, seldom any need for clamps and cable at each and every pipe entering or leaving a domestic hot water cylinder.

Tests will establish whether additional conductors are necessary or not.

Whilst Regulation 514.13.1 requires a warning label to be fixed at or near the point 514.13.1
of connection of every bonding conductor to an extraneous-conductive-part, where there are a number of such connections in close proximity it may be overdoing it to label each one separately, in which case one or perhaps two suitably placed labels could suffice.

Protective multiple earthing 14

14.1 Introduction

The Electricity Safety, Quality and Continuity Regulations 2002 permit the distributor to combine neutral and protective functions in a single conductor provided that, in addition to the neutral to Earth connection at the supply transformer, there are one or more other connections with Earth. The notes of guidance to the Electricity Safety, Quality and Continuity Regulations refer to Electricity Association Limited (now Energy Networks Association) publication G12/3 1995 for details of suitable earthing arrangements. Contact details for the Energy Networks Association are given under Acknowledgements.

The supply neutral may then be used to connect circuit protective conductors of the customer's installation with Earth if the customer's installation meets the requirements of BS 7671. This protective multiple earthing (PME) has been almost universally adopted by distributors in the UK as an effective and reliable method of providing their customers with an earth connection. Such a supply system is described in BS 7671 as TN-C-S (Figure 14.1).

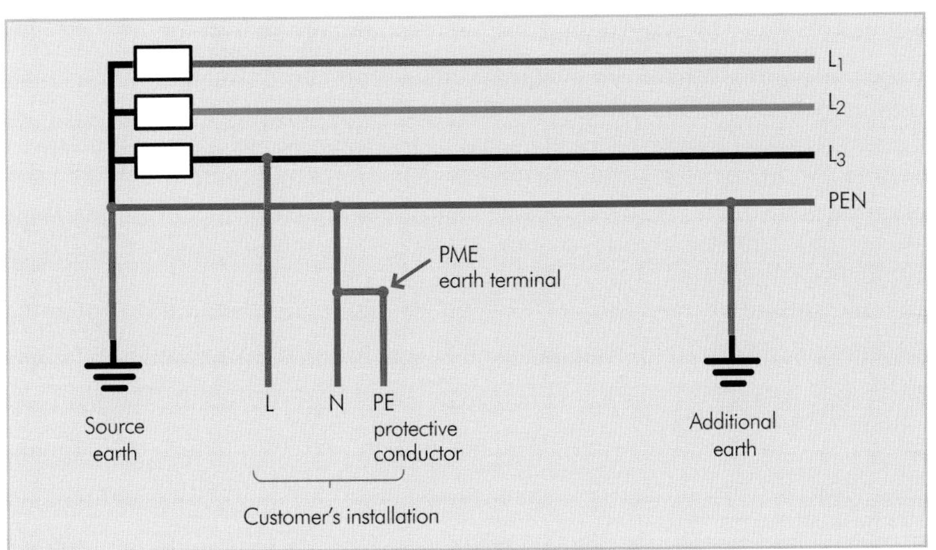

▼ **Figure 14.1**
TN-C-S system

Part 2
Fig 2.4

A protective multiple earthing terminal provides an effective and reliable facility for the majority of installations. However, under certain supply system fault conditions (external to the installation) a potential can develop between the conductive parts connected to the PME earth terminal and the general mass of Earth (see Figure 14.2).

14.2 Supply system

There are multiple earthing points on the supply network, and providing bonding within the building complies with BS 7671 it is unlikely that such a potential as described above would in itself constitute a hazard. However, there are areas of additional risk within or outside buildings, and there are particular situations and installations where it is appropriate to take additional measures for part or all of the installation, such as an additional connection with Earth at the consumer's earth terminal (Figure 14.2). Alternatively, it may be appropriate not to use the PME earthing terminal and provide earth fault protection by means of a separate earth electrode and RCD.

14.3 Potential difference within buildings

The potential difference between true Earth and the PME earth terminal is of importance when

i body contact resistance is low (little clothing, damp/wet conditions), and/or
ii there is relatively good contact with true Earth.

14.4 Potential difference outside buildings

Contact with Earth is always possible outside a building, and, if exposed-conductive-parts and/or extraneous-conductive-parts connected to the PME earth terminal are accessible outside the building, people may be subjected to a voltage difference appearing between these parts and Earth.

14.5 Additional earth electrode for PME supplies

411.4.2 Note In the unlikely event of the PEN conductor of the supply becoming open circuit, touch voltages perhaps causing some discomfort and possibly giving rise to a perceived electric shock may arise on exposed metalwork in customers' installations downstream of the open circuit. The effect can be mitigated by connection of a suitable earth electrode to the main earthing terminal of the customer's installation. The value of the resistance-to-Earth necessary to limit the touch voltages to a given value depends on the load and the network parameters; see Figure 14.2.

Neglecting R_e as it is small compared with R_L and R_A, and neglecting R_B as this errs on the safe side, maximum resistance to Earth, R_A, of the electrode to keep the touch voltage below a given value V_p is

$$R_A = R_L \times \frac{V_p}{(V_s - V_p)}$$

Table 14.1 gives the maximum values of R_A for the additional earth electrode necessary to reduce the touch voltage to 50 V and 100 V, for a range of single-phase loads and a nominal supply voltage of 230 V.

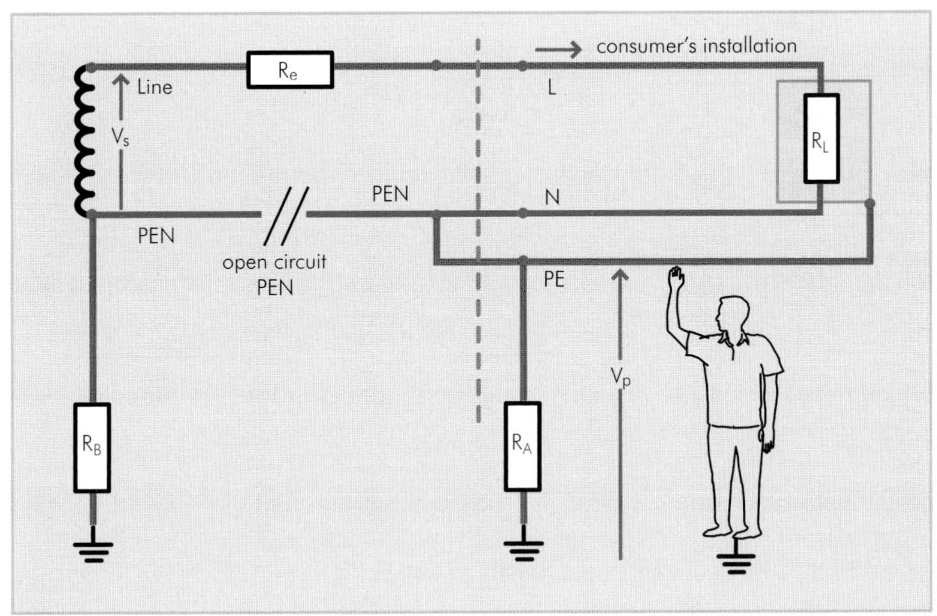

Figure 14.2 Schematic of a PME system with an open circuit PEN conductor

Notes:

V_S is the nominal supply (source) voltage
V_P is the touch voltage
R_e is the external phase supply resistance
R_L is the load resistance (V_S^2 / wattage)
R_A is the resistance of the additional earth electrode including any parallel earths (e.g. water and gas pipes)
R_B is the resistance to Earth of the neutral point of the power supply.

Table 14.1 Additional electrode maximum resistance to Earth, R_A, necessary to reduce the touch voltage to 50 V and 100 V

Load (kW)	R_L (ohms)	R_A (ohms)	
		V_P = 50 V	V_P = 100 V
7	8.2	2.1	5.8
3	19.2	5.1	13.7
2	28.8	7.5	20.5
1	57.6	15.1	41.0

14.6 Special locations

The Energy Networks Association provides guidance on PME in its Engineering Recommendation G12/3, *Requirements for the application of protective multiple earthing to low voltage networks*. The guidance given in this section is based on Part 6 of G12/3.

14.6.1 Remote supplies

Consider a physically isolated building which is supplied via both a long low voltage distribution main and a long individual service cable. If these circumstances are coupled with a substantial load, a potential difference will be evident between the PME terminal and true Earth. This rise in neutral potential is transferred to metalwork connected to the PME terminal, and may be noticeable in such locations as shower areas, where simultaneous contact with true Earth (a wet concrete floor) and the PME terminal (shower pipes) may be possible.

It is recommended that such floors should incorporate a metal grid connected to supplementary bonding within the shower/bathing area. Certainly, if there is a grid in the floor it must be connected to the local supplementary bonding. An alternative approach would be not to use the PME earthing terminal and to afford protection by means of one or more RCDs and an independent earth electrode, that is, treat the shower/bathing area as an installation forming part of a TT system.

14.6.2 Sports pavilions

G12/3 advises that where no shower area exists nor is likely to exist in a sports pavilion, PME may be offered by the distributor provided the appropriate metalwork is bonded. Where a shower exists, PME should only be considered where there is an earth grid in the floor of the shower area which is connected by supplementary bonding conductors to accessible metal pipework, etc.

Alternatives would be to:

i install an additional earth electrode as per 14.5
ii plumb the installation in plastic pipes.

14.6.3 Swimming pools

G12/3 advises that swimming pools supplied with their own dedicated service should have fault protection provided by an RCD. All metalwork should be bonded and connected to an earth electrode. Where a swimming pool forms part of a residence, all metalwork and pipes supplying the pool should be connected to an earth electrode and segregated from the rest of the building. An RCD should then be used to protect the supplies to the pool area and the swimming pool installation treated as part of a TT system.

Where segregation of pipes and metalwork around a pool is impracticable, e.g. in an indoor pool, the installation of a metal grid around a pool and the bonding of adjacent metalwork is recommended together with an RCD in addition to PME.

702.410.3.4.3 The note to Regulation 702.410.3.4.3 advises:

> *Where the supply to the swimming pool is part of a TN-C-S system it is recommended that an earth mat or earth electrode of suitably low resistance, e.g. 20 ohms or less, be installed and connected to the protective equipotential bonding.*

This is the addition of an electrode as advised in 14.5.

14.6.4 Farms and horticultural premises

G12/3 advises that where in remote buildings all extraneous-conductive-parts cannot be bonded to the earthing terminal the pipes and metalwork of isolated buildings, whether or not they have an electricity supply, should be segregated from metalwork connected to the PME earthing terminal. Any supplies to such buildings should be controlled by an RCD and the associated earth electrode and protective conductor should be segregated from any metalwork connected to the PME earthing terminal.

Where segregation is not possible then the alternative of using suitable earth electrodes and RCDs for the whole of the installation should be considered. Alternatively, if a dedicated transformer is used to supply the premises then protective neutral bonding (PNB) may be used; see Figure 14.3. The transformer/system earth is installed close to the consumer's installation.

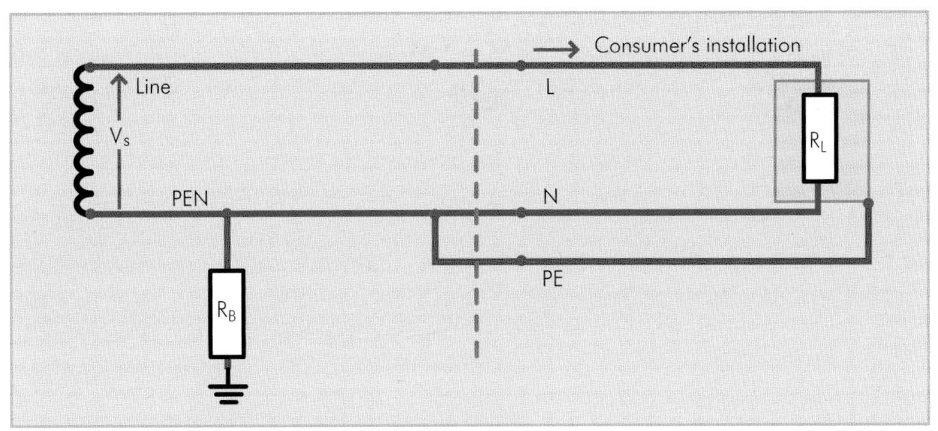

Figure 14.3 Protective neutral bonding (PNB) for installation with a dedicated transformer

14.6.5 Livestock housing (e.g. dairies)

G12/3 advises that particular care must be taken in areas where livestock are housed, as they are sensitive to very small voltages. A suitable metallic mesh must be installed in the concrete bed of a dairy and bonded in accordance with the PME requirements.

If PME is to be applied to an existing dairy the steel reinforcement in the floor should be bonded. Alternatively, if small voltage differences are unacceptable the area concerned should be protected by an RCD and the associated earthing system segregated electrically from the remainder of the installation.

If PME is to be used and the steel reinforcing mesh of the concrete cannot be bonded or does not exist the customer must be advised that, in the case of dairies, the small voltage differences referred to above may adversely affect livestock feeding at milking and also milk output. The note to Regulation 705.415.2.1 of BS 7671 advises that where a metal grid is not laid in the floor a TN-C-S supply is not recommended. 705.415.2.1 Note

14.6.6 Construction sites

It is usually impracticable to comply with the bonding requirements of the Electricity Safety, Quality and Continuity Regulations on construction sites, and a PME earthing terminal should not be provided. Note also that BS 7671 states that a TN-C-S system shall not be used for the supply to a construction site except for the supply to fixed buildings of the construction site. 704.411.3.1

For larger site supplies which require their own substation it will usually be possible to provide an earthing terminal connected directly to the transformer neutral to create a TN-S system. For detailed requirements BS 7375 *Code of Practice for Distribution of Electricity on Construction and Building Sites* and BS 7671 should be consulted.

14.6.7 Caravan parks and marinas

Regulation 9(4) of the Electricity Safety, Quality and Continuity Regulations does not allow a combined neutral and protective conductor to be connected to any metalwork in a caravan or boat. However, this does not preclude a PME earthing terminal being provided for use in permanent buildings at the location, such as the site owner's living premises and any bars or shops. Due to the higher probability of persons being barefooted on caravan sites, the extension of PME earthing to toilet and amenity blocks is not recommended. Supplies to caravans and boats should be two-wire line and neutral supplied through an RCD which must be provided by the customer or site owner. This method of supply is also recommended for toilet or amenity blocks. An independent earth electrode is required.

14.6.8 Petrol filling stations

G12/3 advises that PME facilities should not be provided in petrol filling areas. The filling station area should be treated as a separate system and not have its circuit protective conductors connected to the supply neutral. Where the filling station is part of a larger site, PME facilities may be provided for permanent buildings such as restaurants and shops provided the filling station area is electrically segregated.

Further information is given in the 'Blue Book', *Guidance for Design, Construction, Modification, Maintenance and Decommissioning of Filling Stations,* March 2005, published jointly by APEA and the Institute of Petroleum, which recommends a TT supply for hazardous areas.

14.6.9 Mines and quarries

A supply taken to an underground shaft, or for use in the production side of a quarry, must have an earthing system which is segregated from any system bonded to the PME terminal. Any supply taken to a permanent building can be given a PME terminal provided the building electrical installation wiring complies with BS 7671. Where a mine or quarry requires a supply both to a permanent building and either an underground shaft or the production side of the quarry, precautions must be taken to ensure that these latter supplies have their earthing system segregated from the PME earth system. Further details are given in the following HSE Legal Series (L series) publications:

- L118 *Health and safety at quarries, Quarries Regulations* 1999. Approved Code of Practice, ISBN: 0717624587.
- L128 *The use of electricity in mines, Electricity at Work Regulations* 1989. Approved Code of Practice, ISBN: 0717620743.

14.6.10 Supply terminating in a separate building

Occasionally, a service will terminate in a position remote from the building it supplies.
544.1.1 In this case the size of the PME bonding in the building supplied must be related to the
Table 54.8 size of the incoming supply cable; see Figure 14.4. If the size of the circuit protective conductor of the cable between the supply intake position and the building supplied is less than that of the PME bonding conductor, a suitable additional conductor will need to be installed.

14.6.11 PME and outside water taps

Under an open circuit supply neutral condition, the potential of an outside water tap will rise above Earth potential. A person coming into contact with the tap could receive an electric shock, which could be severe if that person were barefooted. The probability of these two conditions occurring together is considered to be so small that the use of PME where a metal outside tap exists is not precluded.

It is recommended, however, that a plastic insert be provided in the pipe to the outside water tap or the tap be plumbed with plastic pipe.

▼ **Figure 14.4** Supply terminating in a separate building

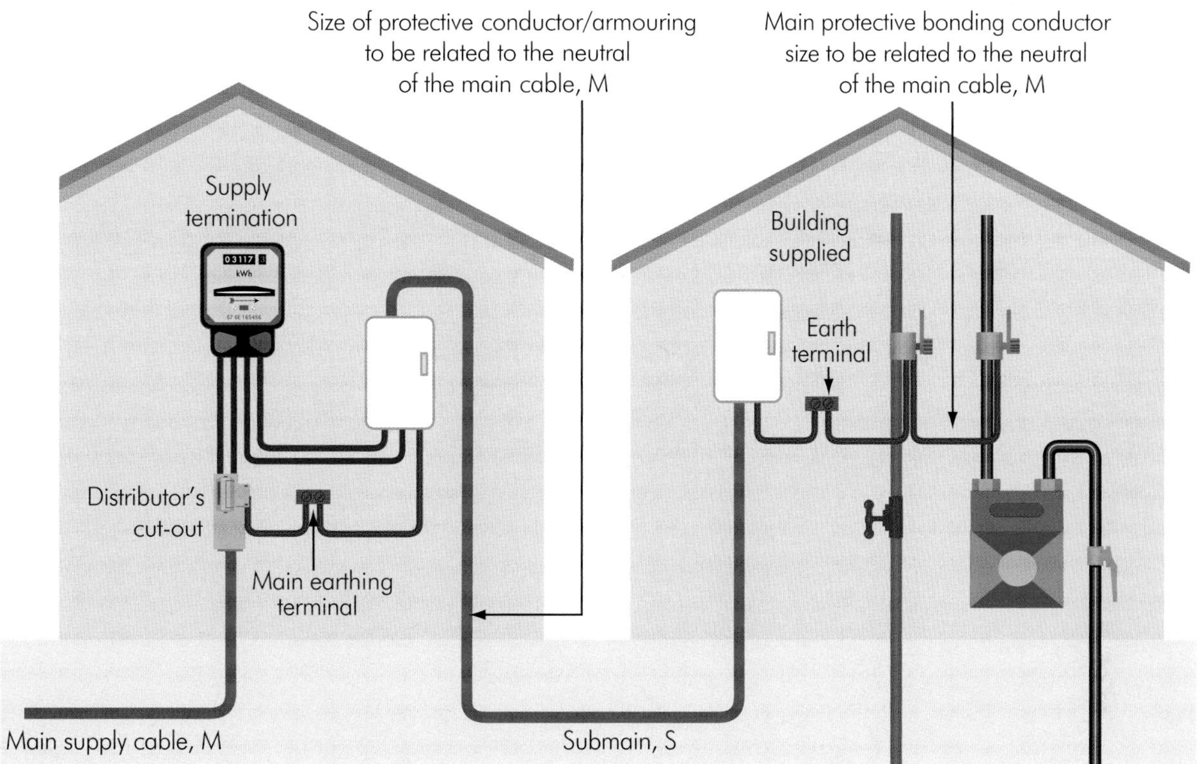

Special installations or locations 15

15.1 The increased risks

The particular requirements set out in the various sections of Part 7 supplement or modify the general requirements of other parts of BS 7671. Supplement here means that additional requirements have to be met. Modify means that restrictions are placed on the choice of protective measures for safety that may be used and/or tighter criteria have to be met. Table 15.1 summarises, for each location, the factors for the supplementary or modified requirements for protection against electric shock. The notes following the table offer further explanation for special installations or locations. Part 7 Sect 700

▼ **Table 15.1** Part 7: The increased risks

Part 7 Section		Wetness[1]	Absence of, or minimal, clothing[2]	Presence of earthed metal[3]	Arduous conditions for equipment[4]	Risk of fire
701	Locations containing a bath or shower	X	X	X		
702	Swimming pools and other basins	X	X	X		
703	Rooms and cabins containing sauna heaters	X	X	X	X	
704	Construction and demolition site installations	X	X	X	X	
705	Agricultural and horticultural premises	X	X	X	X	X
706	Conducting locations with restricted movement[5]	X		X	X	
708	Electrical installations in caravan/camping parks and similar locations	X	X		X	
709	Marinas and similar locations	X	X	X	X	
711	Exhibitions, shows and stands	X		X	X	X
712	Solar photovoltaic (PV) power supply systems[6]					
717	Mobile or transportable units			X	X	
721	Electrical installations in caravans and motor caravans		X	X	X	

continues

▼ **Table 15.1** *continued*

Part 7 Section	Wetness[1]	Absence of, or minimal, clothing[2]	Presence of earthed metal[3]	Arduous conditions for equipment[4]	Risk of fire
740 Temporary electrical installations for structures, amusement devices and booths at fairgrounds, amusement parks and circuses	x		x	x	x
753 Floor and ceiling heating systems[7]					x

Notes:

1 **Wetness.** Wetness means presence of water or humidity, or conditions or work activity causing perspiration. Any of these leads to reduced body resistance by lowering the skin contact resistance. In varying degrees wetness applies to most of the installations and locations listed in Table 15.1.

2 **Absence of, or minimal, clothing.** Apart from bathrooms, swimming pools, saunas, marinas and caravans and their sites, where the absence of clothing or minimal clothing is to be reasonably expected, the situation may occur elsewhere. Summer working conditions on construction sites and in agricultural and horticultural premises, especially greenhouses, may cause persons working there to remove clothing.

3 **Presence of earthed metal.** In most of the locations there is the presence of considerable amounts of earthed metal.

4 **Arduous conditions.** Arduous conditions are environmental conditions, working conditions or activity that give rise to increased risk of electric shock arising from the effect such conditions may have on the installation or equipment supplied by it.

5 **Conducting locations with restricted movement.** Examples of such locations are any metal tank, boiler shell or other metallic vessel into which a person may need to enter, and could also include a large metal pipe that would admit such entry. Such a location does not, however, have to be the inside of a metallic vessel. A plant room could constitute such a location if it is likely that a person working in it will come into bodily contact with substantial areas of conductive material and that such contact is unavoidable.
Here, removal of clothing may be less likely; indeed, special protective clothing may be worn, but perspiration as noted in item **1** above is relevant. If equipment is faulty, the likelihood and severity of electric shock is increased by the conductive environment and the difficulty of escape.

6 **Solar photovoltaic (PV) power supply systems.** The risks include electric shock due to the inability to stop generation when there is daylight, and, during installation, the handling of large and heavy equipment at heights.

7 **Floor and ceiling heating systems.** The risk is the inadvertent penetration of the heating element when performing work such as drilling or nailing. In addition should the element overheat it may present a fire risk.

15.2 Regulation numbering

The regulation numbering of the sections of Part 7 identifies:

- the section
- the chapter, section and regulation amended.

For example, Regulation 702.414.4.5:

- this concerns Section 702 – Swimming pools and other basins, and
- amends Regulation 414.4.5 – regarding basic protection for SELV and PELV.

15.3 Supplementary and modified requirements

The regulations concerned are summarised in Table 15.2. However, it is emphasised that a full study of relevant sections of Part 7 must be made before attempting to apply their particular requirements.

The structure of each section is similar in presentation to the arrangement of the general requirements of BS 7671. Note that each section begins with a definition of its scope of application. It is important to appreciate the scope of any section before attempting to apply its particular requirements.

Table 15.2 Summary of regulations relating to special locations (Note: For a full appreciation of requirements, reference must be made to the applicable regulations.)

Part 7 Section		Locations containing a bath or shower 701	Swimming pools and other basins 702	Rooms and cabins containing sauna heaters 703	Construction and demolition sites 704
Scope and assessment		Scope: 701.1 Zones to be determined: 701.32.1	Scope: 702.11 Zones to be determined: 702.32	Scope: 703.1 Zones to be determined: 703.32	Scope: 704.1 Supplies: 704.313
Protection against electric shock	Prohibited measures	Obstacles, placing out of reach: 701.410.3.5 Non-conducting location, earth-free equipot'l bonding: 701.410.3.6	Obstacles, placing out of reach: 702.410.3.5 Non-conducting location, earth-free equipot'l bonding: 702.410.3.6	Obstacles, placing out of reach: 703.410.3.5 Non-conducting location, earth-free equipot'l bonding: 703.410.3.6	Obstacles, placing out of reach: 704.410.3.5 TN-C-S not permitted (with exception): 704.411.3.1
	Restrictions placed on permitted measures	Electrical separation for a circuit supplying one item of eqpt or one socket-outlet only: 701.413 Where SELV or PELV used, basic protection is required: 701.414.4.5	With exceptions, in zones 0 and 1 only SELV at reduced voltage is permitted: 702.410.3.4 and 702.414.4.5	Where SELV or PELV used, basic protection is required: 703.414.4.5	Where SELV or PELV used, basic protection is required: 704.414.4.5
	Additional protection – RCDs	All circuits: 701.411.3.3	Refer to: 702.410.3.4.1 to 702.410.3.4.3	All circuits: 703.411.3.3 (sauna heater may not need)	Requirement: 704.410.3.10
	Additional protection – supplementary bonding	May be omitted but refer to 701.415.2	All extraneous-conductive-parts: 702.415.2	–	–
Thermal effects	–	–	–	–	–
Selection and erection of equipment	External influences	IP ratings: 701.512.2	IP ratings: 702.512.2	IP ratings: 703.512.2	–
	Wiring systems	–	Restrictions: 702.52	Restrictions: 703.52	Restrictions: 704.52
	Switchgear and controlgear	Restrictions: 701.512.3	Restrictions: 702.53	Restrictions: 703.537.5	Devices for isolation: 704.537.2.2
	Current-using equipment	Restrictions: 701.55	Restrictions: 702.55	–	–
	Plugs and socket-outlets	Not within 3 m of zone 1: 701.512.3	Permitted with restrictions: 702.53	–	Requirement: 704.511.1
	Other equipment	Electric floor heating systems: 701.753	Underwater luminaires: 702.55.2 Eqpt in zone 1: 702.55.4	Heater to comply with BS EN 60335-2-53: 703.55	–

Part 7 Section		Agricultural and horticultural premises 705	Conducting locations with restricted movement 706	Electrical installat'ns in caravan/camping parks and similar locations 708	Marinas and similar locations 709
Scope and assessment		Scope: 705.1	Scope: 706.1	Scope: 708.1	Scope: 709.1 Supplies: 709.313
Protection against electric shock	Prohibited measures	Obstacles, placing out of reach: 705.410.3.5 Non-conducting location, earth-free equipot'l bonding: 705.410.3.6	Obstacles, placing out of reach: 706.410.3.5 –	Obstacles, placing out of reach: 708.410.3.5 Non-conducting location, earth-free equipot'l bonding: 708.410.3.6	Obstacles, placing out of reach: 709.410.3.5 Non-conducting location, earth-free equipot'l bonding: 709.410.3.6
	Restrictions placed on permitted measures	RCDs required: 705.411.1 TN-C not to be used: 705.411.4 Where SELV or PELV used, basic protection is req'd: 705.414.4.5	Additional requirements for the supply to hand-held tools, handlamps and fixed equipment: 706.410.3.10 Additional requirements for ADoS: 706.411 Electrical separation: 706.413 SELV or PELV: 706.414.4	Prohibited to connect a TN-C-S system to a caravan or similar construction: 708.411.4	Prohibited to connect a TN-C-S system to a boat or similar construction: 709.411.4
	Additional protection – RCDs	–		Socket-outlets: 708.553.1.13	Socket-outlets: 709.531.2
	Additional protection – supplementary bonding	Required: 705.415.2.1		–	–
Thermal effects	Protection against fire	Additional requirements: 705.422	–	–	–
Selection and erection of equipment	External influences	IP ratings: 705.512.2	–	IP ratings: 708.512.2	IP ratings and other external influences: 709.512.2
	Accessibility	By livestock: 705.513.2	–	–	–
	Identification	Diagrams: 705.514.9	–	–	–
	Wiring systems	Additional requirements: 705.522	–	Underground, overhead distribution: 708.521.1	Additional requirements: 709.521.1
	Isolation, switching and control	Additional requirements: 705.537	–	–	Additional requirements: 709.537.2
	Protective bonding conductors	705.544.2	–	–	–
	Switchgear and controlgear	–	–	Not more than 20 m: 708.530.3	–
	Current-using equipment	–	–	–	–
	Plugs and socket-outlets	Industrial types: 705.553.1	–	Individual protection: 708.553	Additional requirements: 709.553.1
	Other equipment	Luminaires: 705.559 Safety services: 705.56	–	–	–

Table 15.2 *continued*

Part 7 Section		Exhibitions, shows and stands 711	Solar photovoltaic (PV) power supply systems 712	Mobile or transportable units 717	Electrical installations in caravans and motor caravans 721
Scope and assessment		Scope: 711.1 Supplies: 711.313 External influences: 711.32	Scope: 712.1 Earthing arrangement: 712.3	Scope: 717.1	Scope: 721.1 Supplies: 721.313
Protection against electric shock	Prohibited measures	Obstacles, placing out of reach: 711.410.3.5 Non-conducting location, earth-free equipot'l bonding: 711.410.3.6	D.c. side: 712.410.3.6 Non-conducting location, earth-free equipot'l bonding: 712.410.3.6	Obstacles, placing out of reach: 717.417 Non-conducting location; earth-free equipot'l bonding not recommended: 717.418	Obstacles, placing out of reach: 721.410.3.5 Non-conducting location, earth-free equipot'l bonding: 721.410.3.6
	Restrictions placed on permitted measures	RCD protection: 711.410.3.4 Main bonding: 711.411.3.1.2 TN-C-S prohibited for caravans: 711.411.4 SELV or PELV: 711.414.4.5	D.c. side to be considered energized: 712.410.3 ADoS: 712.411 Double or reinforced insulation: 712.412 SELV or PELV: 712.414.4.5	Automatic disconnection: 717.411	ELV: 721.410.3.3.1 Electrical separation: 721.410.3.3.2 ADoS: 721.411
	Additional protection – RCDs	Socket-outlets: 711.411.3.3	–	Socket-outlets to supply equipment outside unit: 717.415	–
	Additional protection – supplementary bonding	–	–	–	–
Thermal effects	Protection against fire	Heat generation: 711.422	Fault current: 712.434	–	–
Selection and erection of equipment	External influences	–	External influences: 712.512 Accessibility: 712.513	Identification: 717.514	Notices: 721.510 Identification: 721.514 External influences: 721.522
	Wiring systems	Requirements: 711.52	Requirements: 712.522	Requirements: 717.52	Requirements: 721.521
	Switchgear and controlgear	Isolation: 711.537.2	Isolation: 712.537.2	Non-electrical services: 717.528.3	Isolation: 721.537.2
	Current-using equipment	Electric motors: 711.55.4	–	–	–
	Plugs and socket-outlets	711.55.7	–	–	Inlets: 721.55.1
	Other equipment	Luminaires: 711.559	Bonding conductors: 712.54	Requirements: 717.55	Accessories: 721.55.2
Inspection & testing	Upon every assembly	711.6	–	–	–

Part 7 Section		Temporary electrical installations for structures, amusement devices and booths at fairgrounds, amusement parks and circuses 740	Floor and ceiling heating systems 753
Scope and assessment		Scope: 740.1.1 Supplies: 740.313	Scope: 753.1
Protection against electric shock	Prohibited measures	Obstacles not permitted: 740.410.3 Placing out of reach not permitted, except for electric dodgems: 740.410.3 Non-conducting location, earth-free equipot'l bonding: 740.410.3.6	Obstacles, placing out of reach: 753.410.3.5 Electrical separation not permitted: 753.413.1.2 Non-conducting location, earth-free equipot'l bonding: 753.410.3.6
	Restrictions placed on permitted measures	ADoS: 740.411	RCDs and metallic grids: 753.411.3.2
	Additional protection – RCDs	740.415.1	Class II: 753.415.1
	Additional protection – supplementary bonding	740.415.2	–
Thermal effects	Protection against thermal effects and burns	Thermal effects: 740.42	Heating units: 753.424.1 Burns: 753.423
Selection and erection of equipment	External influences	IP rating: 740.512.2	IP rating: 753.512.2
	Wiring systems	740.521	Heating-free areas: 753.52
	Switchgear and controlgear	740.51 740.53	–
	Current-using equipment	–	–
	Connections	740.526	Cold tails: 753.424.1.1 and 753.522.1
	Other equipment	Luminaires: 740.55.1.1 Generators: 740.551 Sockets and plugs: 740.55.7 Supplies: 740.55.8 Dodgems: 740.55.9	
	Identification and notices	–	753.514
Inspection and testing	Upon every assembly	740.6	–

Maximum permissible measured earth fault loop impedance

A

The tables in this appendix provide maximum permissible measured earth fault loop impedances (Z_S) for compliance with BS 7671. The values are those that must not be exceeded when the tests are carried out at an ambient temperature of 10 °C. Table A.5 provides correction factors for other ambient temperatures. 612.9 411.4.6 411.4.7 411.4.8 543.1.3

Where the cables to be used are to Tables 4, 7 or 8 of BS 6004 or Tables 3, 5, 6 or 7 of BS 7211 or are other thermoplastic (PVC) or thermosetting cables to these British Standards, and the cable loading is such that the maximum operating temperature is 70 °C, then Tables A.1 to A.3 give the maximum earth loop impedances for circuits with:

1 protective conductors of copper and having from 1 mm^2 to 16 mm^2 cross-sectional area
2 an overcurrent protective device (i.e. a fuse) to BS 88 Part 2 or Part 6, BS 1361 or BS 3036.

For each type of fuse, two tables are given:

- where the circuit concerned supplies final circuits not exceeding 32 A and the maximum disconnection time for compliance with Regulation 411.3.2.2 is 0.4 s for TN systems, and 411.3.2.2
- where the circuit concerned is a final circuit exceeding 32 A or a distribution circuit and the disconnection time for compliance with Regulation 411.3.2.3 is 5 s for TN systems. 411.3.2.3

In each table the earth fault loop impedances given correspond to the appropriate disconnection time from a comparison of the time/current characteristics of the device concerned and the adiabatic equation given in Regulation 543.1.3. 543.1.3

The tabulated values apply only when the nominal voltage to Earth (U_0) is 230 V.

Table A.4 gives the maximum measured Z_S for circuits protected by circuit-breakers to BS 3871-1 and BS EN 60898, and RCBOs to BS EN 61009.

Note: The impedances tabulated in this appendix are lower than those in Tables 41.2, 41.3 and 41.4 of BS 7671 as the impedances in this appendix are measured values Tables 41.2 to 41.4

at an assumed conductor temperature of 10 °C whilst those in BS 7671 are design figures at the conductor normal operating temperature. The correction factor (divisor) used is 1.24. For smaller section cables the impedance may also be limited by the
543.1.3 adiabatic equation of Regulation 543.1.3. A value of k of 115 from Table 54.3 of BS 7671 is used. This is suitable for PVC insulated and sheathed cables to Tables 4, 7 or 8 of BS 6004 and for thermosetting insulated and sheathed cables to Tables 3, 5, 6 or 7 of BS 7211. The k value is based on both the thermoplastic (PVC) and thermosetting cables operating at a maximum temperature of 70 °C. The IET *Commentary on the Wiring Regulations* provides a full explanation.

▼ **Table A.1** Semi-enclosed fuses. Maximum measured earth fault loop impedance (in ohms) at ambient temperature of 10 °C where the overcurrent protective device is a semi-enclosed fuse to BS 3036

i 0.4 second disconnection (final circuits not exceeding 32 A in TN systems)

Protective conductor (mm^2)	Fuse rating (A)			
	5	15	20	30
1.0	7.7	2.1	1.4	NP
≥ 1.5	7.7	2.1	1.4	0.9

ii 5 seconds disconnection (final circuits exceeding 32 A and distribution circuits in TN systems)

Protective conductor (mm^2)	Fuse rating (A)			
	20	30	45	60
1.0	2.7	NP	NP	NP
1.5	3.1	2.0	NP	NP
2.5	3.1	2.1	1.2	NP
4.0	3.1	2.1	1.3	0.8
≥ 6.0	3.1	2.1	1.3	0.9

Note: NP means that the combination of the protective conductor and the fuse is Not Permitted.

Table A.2 BS 88 fuses. Maximum measured earth fault loop impedance (in ohms) at ambient temperature of 10 °C where the overcurrent protective device is a fuse to BS 88

i 0.4 second disconnection (final circuits not exceeding 32 A in TN systems)

Protective conductor (mm^2)	Fuse rating (A)					
	6	10	16	20	25	32
1.0	6.9	4.1	2.2	1.4	1.2	0.66
1.5	6.9	4.1	2.2	1.4	1.2	0.84
≥ 2.5	6.9	4.1	2.2	1.4	1.2	0.84

ii 5 seconds disconnection (final circuits exceeding 32 A and distribution circuits in TN systems)

Protective conductor (mm^2)	Fuse rating (A)							
	20	25	32	40	50	63	80	100
1.0	1.7	1.2	0.66	NP	NP	NP	NP	NP
1.5	2.3	1.7	1.1	0.64	NP	NP	NP	NP
2.5	2.3	1.8	1.5	0.93	0.55	0.34	NP	NP
4.0	2.3	1.8	1.5	1.1	0.77	0.50	0.23	NP
6.0	2.3	1.8	1.5	1.1	0.84	0.66	0.36	0.22
10.0	2.3	1.8	1.5	1.1	0.84	0.66	0.46	0.33
16.0	2.3	1.8	1.5	1.1	0.84	0.66	0.46	0.34

Note: NP means that the combination of the protective conductor and the fuse is Not Permitted.

▼ **Table A.3** BS 1361 fuses. Maximum measured earth fault loop impedance (in ohms) at ambient temperature of 10 °C where the overcurrent protective device is a semi-enclosed fuse to BS 1361

i 0.4 second disconnection (final circuits not exceeding 32 A in TN systems)

Protective conductor (mm^2)	Fuse rating (A)			
	5	15	20	30
1.0	8.4	2.6	1.4	0.81
1.5	8.4	2.6	1.4	0.93
2.5 to 16	8.4	2.62	1.4	0.93

ii 5 seconds disconnection (final circuits exceeding 32 A and distribution circuits in TN systems)

Protective conductor (mm^2)	Fuse rating (A)					
	20	30	45	60	80	100
1.0	1.7	0.81	NP	NP	NP	NP
1.5	2.2	1.2	0.34	NP	NP	NP
2.5	2.3	1.5	0.52	0.21	NP	NP
4.0	2.3	1.5	0.69	0.37	0.22	NP
6.0	2.3	1.5	0.77	0.53	0.30	0.15
10	2.3	1.5	0.77	0.56	0.40	0.22
16	2.3	1.5	0.77	0.56	0.40	0.29

Note: NP means that the combination of the protective conductor and the fuse is Not Permitted.

Table A.4 Circuit-breakers. Maximum measured earth fault loop impedance (in ohms) at ambient temperature of 10 °C where the overcurrent device is a circuit-breaker to BS 3871 or BS EN 60898 or RCBO to BS EN 61009

For 0.1 to 5 second disconnection times: Table 41.3

Circuit-breaker type	Circuit-breaker rating (A)													
	5	6	10	15	16	20	25	30	32	40	45	50	63	100
1	9.27	7.73	4.64	3.09	2.90	2.32	1.85	1.55	1.45	1.16	1.03	0.93	0.74	0.46
2	5.3	4.42	2.65	1.77	1.66	1.32	1.06	0.88	0.83	0.66	0.59	0.53	0.42	0.26
B	7.42	6.18	3.71	2.47	2.32	1.85	1.48	1.24	1.16	0.93	0.82	0.74	0.59	0.37
3&C	3.71	3.09	1.85	1.24	1.16	0.93	0.74	0.62	0.58	0.46	0.41	0.37	0.29	0.19
D	1.85	1.55	0.93	0.62	0.58	0.46	0.37	0.31	0.29	0.23	0.21	0.19	0.15	0.09

Regulation 434.5.2 of BS 7671:2008 requires the protective conductor csa to meet the requirements of BS EN 60898-1,-2 or BS EN 61009-1, or the energy let-through quoted by the manufacturer. Table A.5 gives minimum protective conductor sizes for energy-limiting Class 3 Type B and C devices.

Table A.5 Minimum protective conductor size for Class 3 Type B and C devices

Energy-limiting Class 3 device rating (A)	Fault level (kA)	Protective conductor csa (mm^2)*	
		Type B	Type C
≤16	≤3	1.0	1.5
≤16	≤6	2.5	2.5
16 < A ≤ 32	≤3	1.5	1.5
16 < A ≤ 32	≤6	2.5	2.5
40	≤3	1.5	1.5
40	≤6	2.5	2.5

* For other device types and ratings or higher fault levels consult manufacturers' data. See Regulation 434.5.2 and the IET publication *Commentary on the IEE Wiring Regulations*.

Table A.6 Ambient temperature correction factors

Ambient temperature (°C)	Correction factor (from 10 °C) (notes 1 and 2)
0	0.96
5	0.98
10	1.00
20	1.04
25	1.06
30	1.08

Notes:

1 The correction factor is given by: {1 + 0.004 (ambient temp – 10 °C} where 0.004 is the simplified resistance coefficient per °C at 20 °C given by BS EN 60228 for both copper and aluminium conductors.

2 The factors are different to those of Table B.2 because Table A.6 corrects from 10 °C and Table B.2 from 20 °C.

The ambient correction factor of Table A.6 is applied to the earth fault loop impedances of Tables A.1 to A.4 if the ambient temperature is other than 10 °C.

For example, if the ambient temperature is 25 °C the measured earth fault loop impedance of a circuit protected by a 32 A type B circuit-breaker to BS EN 60898 should not exceed $1.16 \times 1.06 = 1.23\ \Omega$.

Resistance of copper and aluminium conductors

B

To check compliance with Regulation 434.5.2 and/or Regulation 543.1.3, i.e. to evaluate the equation $S^2 = I^2.t/k^2$, it is necessary to establish the impedances of the circuit conductors to determine the fault current I and hence the protective device disconnection time t. 434.5.2 543.1.3

Fault current $I = U_0/Z_s$

where:

U_0 is the nominal voltage to Earth
Z_s is the earth fault loop impedance

and

$Z_s = Z_e + R_1 + R_2$

where:

Z_e is that part of the earth fault loop impedance external to the circuit concerned
R_1 is the resistance of the line conductor from the origin of the circuit to the point of utilisation
R_2 is the resistance of the protective conductor from the origin of the circuit to the point of utilisation.

Similarly, in order to design circuits for compliance with BS 7671 limiting values of earth fault loop impedance given in Tables 41.2, 41.3 and 41.4 of BS 7671, it is necessary to establish the relevant impedances of the circuit conductors concerned at their operating temperature.

Table B.1 gives values of $(R_1 + R_2)$ per metre for various combinations of conductors up to and including 50 mm² cross-sectional area. It also gives values of resistance (milliohms) per metre for each size of conductor. These values are at 20 °C.

▼ **Table B.1** Values of resistance/metre or $(R_1 + R_2)$/metre for copper and aluminium conductors at 20 °C

Cross-sectional area (mm²)		Resistance/metre or $(R_1 + R_2)$/metre (mΩ/m)	
Line conductor	**Protective conductor**	**Copper**	**Aluminium**
1	–	18.10	
1	1	36.20	
1.5	–	12.10	
1.5	1	30.20	
1.5	1.5	24.20	
2.5	–	7.41	
2.5	1	25.51	
2.5	1.5	19.51	
2.5	2.5	14.82	
4	–	4.61	
4	1.5	16.71	
4	2.5	12.02	
4	4	9.22	
6	–	3.08	
6	2.5	10.49	
6	4	7.69	
6	6	6.16	
10	–	1.83	
10	4	6.44	
10	6	4.91	
10	10	3.66	
16	–	1.15	1.91
16	6	4.23	–
16	10	2.98	–
16	16	2.30	3.82
25	–	0.727	1.20
25	10	2.557	–
25	16	1.877	–
25	25	1.454	2.40
35	–	0.524	0.87
35	16	1.674	2.78
35	25	1.251	2.07
35	35	1.048	1.74
50	–	0.387	0.64
50	25	1.114	1.84
50	35	0.911	1.51
50	50	0.774	1.28

▼ **Table B.2** Ambient temperature multipliers to Table B.1

Expected ambient temperature (°C)	Correction factor*
5	0.94
10	0.96
15	0.98
20	1.00
25	1.02

* The correction factor is given by: {1 + 0.004 (ambient temp – 20 °C)} where 0.004 is the simplified resistance coefficient per °C at 20 °C given by BS EN 60228 for copper and aluminium conductors.

Verification

For verification purposes the designer will need to give the values of the line and circuit protective conductor resistances at the ambient temperature expected during the tests. This may be different from the reference temperature of 20 °C used for Table B.1. The correction factors in Table B.2 may be applied to the Table B.1 values as multipliers to take account of the ambient temperature (for test purposes only).

Multipliers for conductor operating temperature

Table B.3 gives the multipliers to be applied to the values given in Table B.1 for the purpose of calculating the resistance at maximum operating temperature of the line conductors and/or circuit protective conductors in order to determine compliance with, as applicable, the earth fault loop impedance of Table 41.2, 41.3 or 41.4 of BS 7671. Tables 41.2 to 41.4

Where it is known that the actual operating temperature under normal load is less than the maximum permissible value for the type of cable insulation concerned (as given in the BS 7671 tables of current-carrying capacity), the multipliers given in Table B.3 may be reduced accordingly. Appx 4

Table B.3 Multipliers to be applied to Table B.1 to calculate conductor resistance at maximum operating temperature for standard devices (note 4)

Conductor installation	Conductor insulation		
	70 °C Thermoplastic (PVC)	90 °C Thermoplastic (PVC)	90 °C Thermosetting
Not incorporated in a cable and not bunched (note 1)	1.04	1.04	1.04
Incorporated in a cable or bunched (note 2)	1.20	1.28	1.28

Notes:

Table 54.2 **1** See Table 54.2 of BS 7671, which applies where the protective conductor is not incorporated or bunched with cables, or for bare protective conductors in contact with cable covering.

Table 54.3 **2** See Table 54.3 of BS 7671, which applies where the protective conductor is a core in a cable or is bunched with cables.

3 The multipliers given in Table B.3 for both copper and aluminium conductors are based on a simplification of the formula given in BS EN 60228, namely that the resistance-temperature coefficient is 0.004 per deg C at 20 °C.

4 Standard devices are those described in Appendix 3 of BS 7671 (fuses to BS 1361, BS 88, BS 3036, circuit-breakers to BS EN 60898 types B, C, and D) and BS 3871-1.

Index

C

D

E

Z

IEE Wiring Regulations and associated publications

The IEE prepares regulations for the safety of electrical installations for buildings, the *IEE Wiring Regulations* (BS 7671 *Requirements for Electrical Installations*), which have now become the standard for the UK and many other countries. It also recommends, internationally, the requirements for ships and offshore installations. The IEE provides guidance on the application of the installation regulations through publications focused on the various activities from design of the installation through to final test and then maintenance. This includes a series of eight Guidance Notes, two Codes of Practice and Model Forms for use in Wiring Installations.

Requirements for Electrical Installations BS 7671:2008 (IEE Wiring Regulations, 17th Edition)
Order book PWR1700B Paperback 2008
ISBN: 978-0-86341-844-0 **£75**

On-Site Guide (BS 7671:2008 17th Edition)
Order book PWGO170B 188pp Paperback 2008
ISBN: 978-0-86341-854-9 **£22**

Wiring Matters Magazine **FREE**
If you wish to receive a FREE copy or advertise in Wiring Matters please visit www.theiet.org/wm

IEE Guidance Notes

A series of Guidance Notes has been issued, each of which enlarges upon and amplifies the particular requirements of a part of the IEE Wiring Regulations.

Guidance Note 1: Selection & Erection of Equipment, 5th Edition
Order book PWG1170B 216pp Paperback 2009
ISBN: 978-0-86341-855-6 **£30**

Guidance Note 2: Isolation & Switching, 5th Edition
Order book PWG2170B 88pp Paperback 2009
ISBN: 978-0-86341-856-3 **£25**

Guidance Note 3: Inspection & Testing, 5th Edition
Order book PWG3170B 128pp Paperback 2008
ISBN: 978-0-86341-857-0 **£25**

Guidance Note 4: Protection Against Fire, 5th Edition
Order book PWG4170B 104pp Paperback 2009
ISBN: 978-0-86341-858-7 **£25**

Guidance Note 5: Protection Against Electric Shock, 5th Edition
Order book PWG5170B 144pp Paperback 2009
ISBN: 978-0-86341-859-4 **£25**

Guidance Note 6: Protection Against Overcurrent, 5th Edition
Order book PWG6170B 104pp Paperback 2009
ISBN: 978-0-86341-860-0 **£25**

Guidance Note 7: Special Locations, 3rd Edition
Order book PWG7170B 144pp Paperback 2009
ISBN: 978-0-86341-861-7 **£25**

Guidance Note 8: Earthing & Bonding, 1st Edition
Order book PWRG0241 168pp Paperback 2007
ISBN: 978-0-86341-616-3 **£25**

continues overleaf ▶

Other guidance publications

Commentary on IEE Wiring Regulations (17th Edition, BS 7671:2008)
Order book PWR08640
c.432pp Hardback 2009
ISBN: 978-0-86341-966-9 **£65**

Electrical Maintenance, 2nd Edition
Order book PWR05100
228pp Paperback 2006
ISBN: 978-0-86341-563-0 **£40**

Code of Practice for In-service Inspection and Testing of Electrical Equipment, 3rd Edition
Order book PWR08630
152pp Paperback 2007
ISBN: 978-0-86341-833-4 **£40**

Electrical Craft Principles, Volume 1, 5th Edition
Order book PBNS0330
344pp Paperback 2009
ISBN: 978-0-86341-932-4 **£25**

Electrical Craft Principles, Volume 2, 5th Edition
Order book PBNS0340
432pp Paperback 2009
ISBN: 978-0-86341-933-1 **£25**

Electrician's Guide to the Building Regulations, 2nd Edition
Order book PWGP170B
234pp Paperback 2008
ISBN: 978-0-86341-862-4 **£22**

Electrical Installation Design Guide: Calculations for Electricians and Designers
Order book PWR05030
186pp Paperback 2008
ISBN: 978-0-86341-550-0 **£22**

Electrician's Guide to Emergency Lighting
Order book PWR05020
88pp Paperback 2009
ISBN: 978-0-86341-551-7 **£22**

Electrical training courses

We offer a comprehensive range of technical training at many levels, serving your training and career development requirements as and when they arise.

Courses range from Electrical Basics to Qualifying City & Guilds or EAL awards.

Train to the 17th Edition BS 7671:2008

- Update from 16th to 17th Edition
- Understand the changes
- New qualifying awards C&G/EAL
- Meet industry standards

Qualifying Courses

- Certificate of Competence Management of Electrical Equipment Maintenance (PAT) – 1 day
- Certificate of Competence for the Inspection and Testing of Electrical Equipment (PAT) – 1 day
- Certificate in the Requirements for Electrical Installations – 3 days
- Upgrade from 16th Edition achieved since 2001 – 1 day
- Certificate in Fundamental Inspection, Testing and Internal Verification – 3 days
- Certificate in Inspection, Testing and Certification of Electrical Installations – 3 days

Other 17th Edition Courses

- Earthing & Bonding – For designers and electrical contractors who require a good working knowledge of the E & B arrangements as required by BS 7671:2008
- 17th Edition Design – BS 7671 and the principles associated with the design of electrical installations

To view all our current courses and book online, visit **www.theiet.org/coursesbr**

To discuss your training requirements and for on-site group training, please speak to one of our advisors on +44 (0)1438 767289

For more information, visit www.theiet.org/wiringregs

Order Form

How to order

BY PHONE:
+44 (0)1438 767328
BY FAX:
+44 (0)1438 767375
BY EMAIL:
sales@theiet.org
BY POST:
The Institution of Engineering and Technology, PO Box 96, Stevenage SG1 2SD, UK
OVER THE WEB:
www.theiet.org/books

*Postage/Handling: Postage within the UK is £3.50 for any number of titles. Outside UK (Europe) add £5.00 for first title and £2.00 for each additional book. Rest of World add £7.50 for the first book and £2.00 for each additional book. Books will be sent via air-mail. Courier rates are available on request, please call +44 (0) 1438 767328 or email sales@theiet.org for rates.

** To qualify for discounts, member orders must be placed directly with the IET.

GUARANTEED RIGHT OF RETURN:
If at all unsatisfied, you may return book(s) in new condition within 30 days for a full refund. Please include a copy of the invoice.

DATA PROTECTION:
The information that you provide to the IET will be used to ensure we provide you with products and services that best meet your needs. This may include the promotion of specific IET products and services by post and/or electronic means. By providing us with your email address and/or mobile telephone number you agree that we may contact you by electronic means. You can change this preference at any time by visiting www.theiet.org/my.

Details

Name:

Job Title:

Company/Institution:

Address:

Postcode: Country:

Tel: Fax:

Email:

Membership No (if Institution member):

Payment methods

☐ By **cheque** made payable to The Institution of Engineering and Technology

☐ By **credit/debit card:**

☐ Visa ☐ Mastercard ☐ American Express ☐ Maestro Issue No:____________

Valid from: ☐☐ ☐☐ Expiry Date: ☐☐ ☐☐ Card Security Code: ☐☐☐☐
(3 or 4 digits on reverse of card)

Card No: ☐☐☐☐ ☐☐☐☐ ☐☐☐☐ ☐☐☐☐

Signature____________________ Date ____________
(Orders not valid unless signed)

Cardholder Name:

Cardholder Address:

Town: Postcode:

Country:

☐ By official **company purchase order** (please attach copy)

EU VAT number:____________________

Ordering information

Quantity	Book No.	Title/Author	Price (£)
		Subtotal	
		- Member discount**	
		+ Postage /Handling*	
		+ VAT (if applicable)	
		Total	

Membership

Passionate about engineering? Committed to your career?

Do you want to join an organisation that is inspiring, insightful and innovative?

One of the most highly recognised knowledge sharing networks in the world, membership to the Institution of Engineering and Technology (IET) is for engineers and technologists working or studying in an increasingly multidisciplined, digital and global environment.

Joining the IET and having access to tailored products and services will become invaluable for your career and can be your first step towards professional qualifications.

You could take advantage of ...

- 18 issues per year of the industry's leading publication, *E&T* magazine.
- Professional development and career support services to help gain professional registration.
- Discounted rates on dedicated training courses, seminars and events covering a wide range of subjects and skills.
- Watch live IET.tv event footage at your desktop via the internet, ask the speaker questions during live streaming and feel part of the audience without physically being there.
- Access to over 100 local networks around the world.
- Meet like-minded professionals through our array of specialist online communities.
- Instant online access to over 70,000 books, 3,000 periodicals and full-text collections of electronic articles – wherever you are in the world.
- Discounted rates on IET books and technical proceedings.

Join online today at www.theiet.org/join or contact our membership and customer service centre on +44 (0)1438 765678

Professional Registration

What type of registration is for you?

Engineering Technicians (EngTech) are involved in applying proven techniques and procedures to the solution of practical engineering problems. You will carry supervisory or technical responsibility, and are competent to exercise creative aptitudes and skills within defined fields of technology. Engineering Technicians also contribute to the design, development, manufacture, commissioning, operation or maintenance of products, equipment, processes or services.

Incorporated Engineers (IEng) maintain and manage applications of current and developing technology, and may undertake engineering design, development, manufacture, construction and operation. Incorporated Engineers are engaged in technical and commercial management and possess effective interpersonal skills.

Chartered Engineers (CEng) develop appropriate solutions to engineering problems, using new or existing technologies, through innovation, creativity and change. They might develop and apply new technologies, promote advanced designs and design methods, introduce new and more efficient production techniques, marketing and construction concepts, pioneer new engineering services and management methods. Chartered Engineers are engaged in technical and commercial leadership and possess interpersonal skills.

For further information on Professional Registration (EngTech/IEng/CEng), tel: +44 (0)1438 765673 or email: membership@theiet.org